# Welcome to Scientifica 8

Text © Peter Ellis, Lawrie Ryan, David Sang, Jane Taylor 2005
Original illustrations © Nelson Thornes Ltd 2005

The right of Peter Ellis, Lawrie Ryan, David Sang, Jane Taylor to be identified as authors of this work has been asserted by them in accordance with the Copyright, Designs and Patents Act 1988.

Published in 2005 by:
Nelson Thornes Ltd
Delta Place
27 Bath Road
CHELTENHAM
GL53 7TH
United Kingdom

05 06 07 08 09 /
10 9 8 7 6 5 4 3 2

A catalogue record for this book is available from the British Library

ISBN 0 7487 9185 X

Illustrations by Mark Draisey, Ian West, Bede Illustration
Cover illustration by Andy Parker
Page make-up by Wearset Ltd

Printed and bound in China by Midas

## Introduction

This workbook contains the really essential information you need to help you do well in Year 8. The work is split into short chunks within each of the 12 Units. You can see the names of the units in the contents list.

There are regular 'CHECKPOINT' questions to make sure you understand ideas as you meet them. There are also further homework tasks to try in which you can apply your learning. These are clearly linked to each section of the Unit you are studying. You'll find loads of different styles of exercise – so you won't get bored with Scientifica!

This workbook is linked to the exciting new Scientifica textbooks. You will find a lot more detail in these textbooks, including practical activities.

However, you can use these 'write-on' books separately and still be sure that you are well prepared for your tests at the end of Key Stage 3.

**Enjoy your journey through science with Scientifica!**

## CONTENTS

#  Food and digestion

## 8A1 Food

Food gives us **nutrients** that are essential to health. We need large amounts of three types of nutrient – **carbohydrates**, **proteins** and **fats**.

Carbohydrates come in two forms – **sugars** and **starches**.

Each day we also need:

- About 1.5 litres of water
- Vitamins
- Minerals
- Fibre; we cannot digest fibre, but it is important for health.

| Nutrient | What we use it for | Where we can get it from |
|---|---|---|
| carbohydrate | energy for activities | potatoes, pasta, rice, bread, sugar |
| protein | growth, making new cells | meat, fish, nuts, cheese, eggs, beans, peas, lentils |
| fat | storing energy, making new cells, keeping warm | butter, nuts, vegetable oils, red meat, salmon, cheese |

### Testing carbohydrates

- Starch makes iodine turn blue-black.
- Benedict's solution turns red when warmed with glucose (a sugar).

### CHECKPOINT

1 List the foods in the picture.

_____

_____

_____

_____

_____

2 What types of nutrient are they?

_____

3 What are each of the nutrients used for?

_____

_____

## 8A2 Vitamins and minerals

If we had no vitamins and minerals in our food, we would suffer from diseases such as scurvy, beriberi and rickets. These are called **deficiency diseases**. We need the **Recommended Daily Amount (RDA)** to stay healthy. Vitamins and minerals are usually found in fresh food.

We limey sailors are never scurvy dogs

Without Vitamin C, sailors became tired, they bruised easily and their gums bled.

| Vitamin | Use | Food source |
|---|---|---|
| A | eyesight, healthy skin | milk, oily fish, carrots, cheese |
| B1 (thiamine) | healthy nerves, release of energy | wholegrain cereals, yeast, eggs, liver, beans |
| C (ascorbic acid) | processes in cells, absorbing iron | citrus fruits, salad greens, blackcurrants |
| D | bone growth | oily fish, milk, eggs, cheese |

| Mineral | Use | Food source |
|---|---|---|
| calcium | bones, teeth, nerves | cheese, milk, bread, spinach, sardines |
| iron | red blood cells, releasing energy | meat, eggs, cereals, apricots, spinach |

### CHECKPOINT

Look back at the photograph in 8A1.

1 What vitamins and minerals are in the food shown?

_____

2 How could your health be affected by the missing vitamins?

_____

# 8A3 Balanced diet

To stay healthy we must have the correct amounts of each type of nutrient in our diets.

## Energy

Energy comes mainly from carbohydrates and fats. We use energy:

- For life processes, such as breathing and circulation of the blood
- To grow
- To repair damaged tissues
- To move
- To keep warm.

The amount of energy you need depends on the size of your body and how active you are.

- Someone who is very active may need 1000 kJ/day extra.

| Person | Energy needs (kJ/day) |
| --- | --- |
| child | 6800 |
| woman | 9500 |
| man | 11 500 |

- People who don't get enough of a nutrient become **malnourished**.

- If people eat more energy foods than they need, they store it as fat and become **obese**. Obesity can lead to diseases such as diabetes.

- A **sedentary** person has little exercise and needs less energy foods.

◀ CHECKPOINT ▶

Why do these people need more energy foods than average?

A footballer:

_____

An Eskimo:

_____

A pregnant woman:

_____

# 8A4 The digestive system

Digestion breaks down the large molecules of food into small molecules that can be absorbed into the blood. Food passes through these organs:

- **Mouth** – food is taken in (**injested**); teeth break up particles of food; **saliva** coats food and starts to break it up.
- **Oesophagus** – pushes food down to the gut by **peristalsis**.
- **Stomach** – churns food into a paste; makes acid.
- **Small intestine** – gastric juices **digest** food; nutrients are absorbed.
- **Large intestine** – absorbs water.
- **Rectum** – stores waste as **faeces**.
- **Anus** – a ring of muscle that opens when faeces are **ejested**.

The **pancreas** makes digestive juices.

The **liver** makes **bile**, which digest fats, and carries away dead, red blood cells.

**Villi** are tiny, finger-shaped parts of the intestine with lots of blood vessels. Food molecules pass into the blood and are carried to the liver.

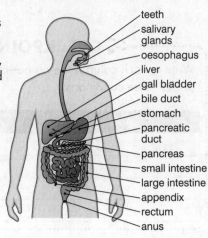

teeth
salivary glands
oesophagus
liver
gall bladder
bile duct
stomach
pancreatic duct
pancreas
small intestine
large intestine
appendix
rectum
anus

◀ CHECKPOINT ▶

1 What is the meaning of these?
Injested:

_____

Digested:

_____

Ejested:

_____

2 Where do they happen?

_____

_____

3 Which organs are involved?

_____

_____

_____

# 8A5 Digesting food

Proteins, fats and carbohydrates such as starch are very large molecules. They are made of many small units joined together.

Digestive juices contain **enzymes**. These are biological catalysts that break down the large molecules into the small molecules that are the **products of digestion**.

The small molecules can pass through the walls of the villi and enter the blood. They are carried to the liver.

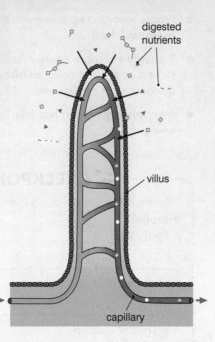

digested nutrients

villus

capillary

| Nutrient | Enzyme | Where it happens | Broken down to |
|---|---|---|---|
| starch | amylase | mouth (saliva) | glucose |
| carbohydrates | carbohydrase | small intestine | glucose |
| proteins | protease | stomach and small intestine | amino acids |
| fats | lipase | small intestine | fatty acids and glycerol |

In the liver, glucose molecules are joined together to form **glycogen**, which stores energy until it is needed.

**◄ CHECKPOINT ►**

In the space below, sketch the digestive system and mark on it where foods are digested and the enzymes used.

# 8A6 Enzymes

- Enzymes are proteins and can be affected by chemicals and heat.

- Enzymes are catalysts that speed up chemical reactions.

- If the temperature is too low, the reaction is slow. If the temperature is too high, the enzyme cannot work – it is **denatured** and food molecules are not broken down.

- In humans, most enzymes work best at 37 °C.

- Acids and alkalis also affect how enzymes work. Each enzyme has an ideal pH.

- Amylase in saliva works best in alkaline solutions.

- Proteases in the stomach work best at pH 2–3, so the stomach makes acid to keep the pH low.

- The small intestine is alkaline to suit the enzymes that work there.

**◄ CHECKPOINT ►**

Write in each column the names of the enzymes that prefer those conditions:

| Acid | Alkali |
|---|---|
| | |
| | |
| | |
| | |

## Food (8A1)

**1** Which type of food is each of the following sentences referring to?

a   It is used to make new cells, nails and hair. _____

b   It insulates the body and is used to make hormones. _____

c   It gives the body energy to move and grow. _____

**2** Molly and Benson tested some foods with iodine.

a   What did they see when iodine was put on potato?

_____

b   What did this tell Molly and Benson?

_____

c   They put some Benedict's solution with a fruit drink and warmed the mixture. It turned red.
What does this show?

_____

**3** For lunch Benson had fish and chips followed by an ice-cream, while Molly had a cheese
sandwich. What nutrients are found in these?

Chips: _____      Fish: _____      Ice-cream: _____

Bread: _____      Cheese: _____

## Vitamins and minerals (8A2)

**1** Fill in the gaps:

a   Frederick Gowland Hopkins fed rats with artificial proteins, carbohydrates and fats. They

died from _____ diseases. He added milk to the diet of other rats and they

stayed _____ . This showed that milk contained _____ and

_____ that the rats needed.

b   British sailors were called 'limeys' after they were given lemons

and _____ to eat. These _____ fruits

contain vitamin _____ (ascorbic acid) that

protected the sailors from the disease _____ .

c   Rickets is a disease, which children get when their food does

not contain vitamin _____ . It is needed to help

the mineral _____ form bones and is found in _____ .

**2** Complete the table:

| Deficiency disease | Vitamin or mineral missing | Foods that would cure the disease |
|---|---|---|
| beriberi | B1 (thiamine) | |
| night blindness | | carrots, oily fish |
| anaemia | iron | |

## Balanced diet (8A3)

**1 a** Unscramble these letters to make words that describe problems caused by a poor diet and write the cause of each:

b o y t i e s _____

r o u n d h i s m a l e _____

**2** Choose a number from the list that is the energy requirements (in kJ/day) for the following people:  12 500   11 500   9500   4200

**A baby:** _____   **A male bus driver:** _____

**A female secretary:** _____   **A male rower:** _____

**3** Pete is a teenager who needs about 9600 kJ/day to be healthy. Every day he eats foods made up of about 300 g of starch, 60 g of sugar and 100 g of fatty foods (starch and sugar provide 17 kJ/g, fats give 38 kJ/g).

**a** Is Pete getting too little, too much or about the right amount of energy foods? _____

**b** State three things Pete should do to stay healthy.

_____

_____

_____

## The digestive system (8A4)

**1 a** Label the diagram with these organs of the digestive system:

**small intestine     stomach**
**anus     mouth     rectum**
**large intestine     oesophagus**

**a** Mark the diagram with these letters showing where they take place:

**A:** water is reabsorbed
**B:** food is churned into a fine paste
**C:** food is absorbed into the blood
**D:** food is pushed along
**E:** food is ingested
**F:** faeces are egested

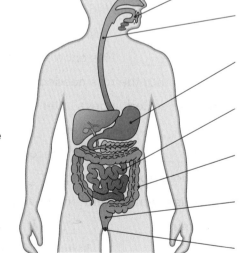

**2** Join each word with its meaning:

| Peristalsis | | Organ that makes digestive juices |
| Pancreas | | Substance made by liver to help digest fats |
| Villi | | The process of pushing food through the gut |
| Bile | | Finger-like parts of the intestine |

## Digesting food (8A5)

**1** Complete the following digestion reactions:

**a** starch ———amylase——→ _____

**b** _____ ———proteases——→ _____

**c** _____ ———————→ _____ + glycerol

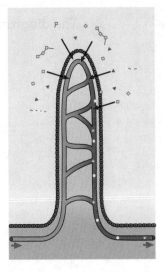

**2** Describe what is happening in the diagram.

_____

_____

**3 a** What is glycogen made from? _____

**b** What is it used for? _____

## Enzymes (8A6)

**1** Unscramble the letters to find the words that complete these sentences:

**n e e d a r u s t    l i k e a l a n    s a l t y c a t s**

**a** Substances that speed up chemical reactions are _____

**b** When a protein, such as an enzyme, is heated it _____

**c** A solution with a pH greater than 7 is _____

**2** Colour the diagram: red to show where enzymes work best in alkaline solution, and blue for acid.

**3** Starch solution was warmed in each test tube. Amylase was added to each. Two minutes later, a drop of iodine was added.

Explain what has happened in each test tube.

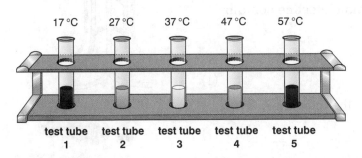

17 °C    27 °C    37 °C    47 °C    57 °C

test tube 1    test tube 2    test tube 3    test tube 4    test tube 5

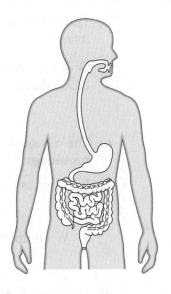

Test tube 1: _____

Test tube 2: _____

Test tube 3: _____

Test tube 4: _____

Test tube 5: _____

# APPLY YOUR KNOWLEDGE

**(SAT-style questions)**

**1** The diagram shows part of the digestive system.

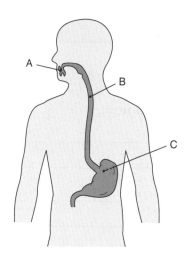

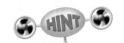

**HINT** Amylase works best in alkaline solutions.

**a)** In the diagram, which letter:

**i)** points to the **oesophagus?** ______

**ii)** points to an organ that contains an acid? ______

**iii)** points to the organ where the enzyme **amylase** is made? ______

**iv)** points to where **peristalsis** happens? ______ (4)

**HINT** Peristalsis is contractions of muscles pushing food along.

**b)** The diagram shows one of the villi in the small intestine:

Give one reason why glucose can pass into the blood through the villi but starch cannot.

______ (1)

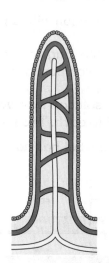

**c)** The digestive juices in the small intestine contain **proteases**:

**i)** What is a protease?

______ (1)

**ii)** What does protease do in the intestine?

______ (1)

**HINT** Think about what stops enzymes working.

**d)** The enzyme lipase breaks down fats into fatty acids and glycerol in the intestine. Explain why this reaction only works well at temperatures close to body temperature (37 °C).

______

______ (1)

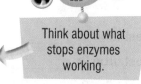

**2** The table on the label of a baked beans can lists the amount of various food types:

| | Amount per 100 g |
|---|---|
| energy | 323 kJ |
| protein | 4.9 g |
| carbohydrate | 13.7 g |
| fat | 0.2 g |
| fibre | 3.9 g |

a) i) Which of the substances listed is not digested by humans, but is needed in a healthy diet?

_____ (1)

HINT

Look at the label and think about the three main groups of nutrients needed.

ii) Give one reason why proteins are needed in our diet.

_____ (1)

iii) Why would a diet made up of just baked beans not be healthy?

_____ (1)

c) Which two types of nutrient provide most of the energy in a normal diet?

1. _____  2. _____ (2)

c) What difference would there be in the diet of a girl who plays tennis regularly and a girl who just watches TV?

_____

(2)

HINT

Watching TV doesn't take much energy!

**3** In the nineteenth century, scientists had found that food was made up of proteins, carbohydrates and fats.

Frederick Gowland Hopkins suggested that other substances were needed for a healthy diet.

He did an experiment:

- He made artificial food from pure proteins, fats and carbohydrates.
- He fed six rats with this artificial food.
- He fed six other rats with the artificial food and a little milk.
- He observed what happened to the rats.

The six that were fed the milk remained healthy. The other rats died.

a) What was Hopkins' conclusion to his experiment?

_____

_____ (2)

HINT

Check what Hopkins suggested before he did the experiment.

b) Why did Hopkins have to make artificial food for his rats?

_____ (1)

c) Small amounts of substances were found in milk, in addition to proteins, carbohydrates and fats.

What are these substances called?

_____ (1)

# **8B** Respiration

## 8B1 Respiration

All living cells need energy. Most cells get their energy from glucose obtained from food. **Aerobic respiration** is a chemical reaction that releases energy by reacting glucose with oxygen.

In animals, blood carries glucose and oxygen to cells:

**glucose + oxygen → carbon dioxide + water + energy**

### ⬤ Detecting respiration

We can show that carbon dioxide gas is given off, by using either:

- Limewater that turns cloudy, or
- Sodium hydrogencarbonate indicator that turns red.

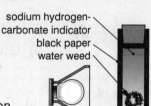

sodium hydrogen-carbonate indicator
black paper
water weed

We need energy from respiration for a number of reasons:

| Use | How the energy is used |
|---|---|
| growth | making new cells |
| synthesis | making the substances that cells need, e.g. proteins |
| keeping temperature constant | mammals and birds keep their bodies warm; humans' body temperature is 37 °C |
| movement | muscles use energy to move |
| active transport | cells take in and remove substances |

If cells have more glucose than they need, it is stored as fat.

### ◀ CHECKPOINT ▶

1  For respiration: what are:
   a) the reactants?

   _____

   b) the products?

   _____

2  How could you prove that an organism was respiring?

   _____

## 8B2 The heart and circulation

Blood carries the substances needed by cells to the cells and takes away the substances that cells produce. The **circulatory system** pumps blood around your body.

Every time your heart beats, it produces a surge of blood called the **pulse**. It can be felt in any artery that is close to the skin.

### ⬤ The heart

The heart is a pump that pushes blood around the circulatory system. Blood enters the left-hand side and is pumped to the lungs. It returns to the right-hand side and then is pumped to all the organs of the body.

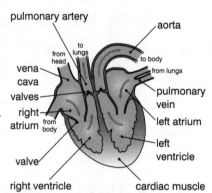

pulmonary artery
aorta
to lungs
from head
to body
from lungs
vena cava
valves
right atrium
from body
pulmonary vein
left atrium
left ventricle
valve
right ventricle
cardiac muscle

### ⬤ Blood vessels

Blood leaves the heart in thick-walled **arteries**. The arteries divide and get smaller. The smallest blood vessels are **capillaries** that carry blood to every cell. The blood is carried back to the heart in thin-walled **veins**.

### ◀ CHECKPOINT ▶

Add arrows to the diagram showing the journey of a drop of blood from the moment it enters the left-hand side of the heart until it reaches the body.

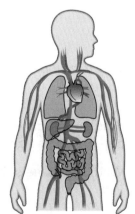

# 8B3 Supplying cells

- Red blood cells contain **haemoglobin**.

- Oxygen joins to haemoglobin to form **oxyhaemoglobin**.

- **Diffusion** is the movement of molecules to places where there is less of the substance.

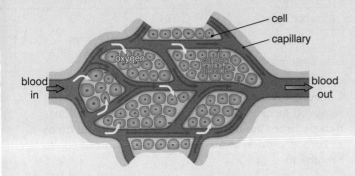

- Glucose, oxygen and other nutrients diffuse from blood into cells, through the walls of the capillaries.

- Carbon dioxide, wastes and other substances diffuse from cells into the blood.

## Anaerobic respiration

If cells do not get enough oxygen, they can release energy from glucose without it. This reaction produces less energy and forms **lactic acid** instead of carbon dioxide. Lactic acid causes pain and cramp in muscles. Oxygen is needed to remove the lactic acid.

Yeast makes **ethanol** in another form of anaerobic respiration, called **fermentation**.

### ◄ CHECKPOINT ►

1 Why does oxygen move from red blood cells to tissue cells?

_____

2 Why do your muscles have to respire anaerobically when you exercise hard?

_____

# 8B4 Breathing

Breathing is how oxygen enters your body and carbon dioxide and water vapour leave it. Moving air in and out of your lungs is called **ventilation**.

## Breathing in: inhaling

- Muscles over the ribs contract and the ribs rise.

- The **diaphragm** is a muscle. When it contracts it flattens. The volume of your chest increases and air is pushed into your lungs.

## Breathing out: exhaling

- Muscles relax.

- The ribs drop and the diaphragm returns to its dome shape. The volume of the chest decreases and pushes air out of the lungs.

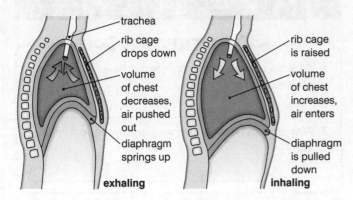

The breathing tubes are held open by rings of **cartilage**.

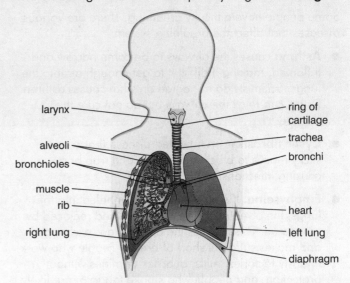

### ◄ CHECKPOINT ►

1 Check the positions of the trachea, larynx, bronchi, bronchioles and alveoli.
2 Describe the journey of a molecule of oxygen from the air to the lungs.

_____

_____

_____

# 8B5 What happens in your lungs

**Gas exchange** takes place in the alveoli. Alveoli have very thin walls, but a very large surface area. They are surrounded by lots of capillaries. Gases can diffuse each way through the walls of the alveoli.

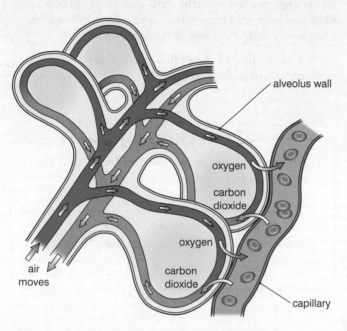

The composition of exhaled air is a little different to that of inhaled air:

| Gas | In inhaled air (%) | In exhaled air (%) |
|---|---|---|
| oxygen | 21 | 18 |
| carbon dioxide | 0.04 | 3 |
| nitrogen | 78 | 78 |
| other gases | 1 | 1 |

Carbon dioxide makes blood acidic so it has to be removed quickly. Removing wastes, such as carbon dioxide, from the body is called **excretion**. The amount of carbon dioxide in the blood controls how often we breathe in.

### ◄ CHECKPOINT ►

Why is it important that the alveoli have a large surface area?

_____

# 8B6 Lung damage

Some people have difficulty breathing. There are various diseases that affect the breathing system:

- **Asthma** causes the airways to be come narrow and inflamed, making it difficult to get enough air into the lungs. Scientists do not agree on what causes asthma. Drugs can relax the airways but do not cure the disease.

- **Cystic fibrosis** is an inherited disease that causes sticky mucus to block the airways and trap bacteria causing infections.

- **Emphysema.** If smoke or dust are inhaled into the lungs, the alveoli become damaged and replaced by scar tissue. This reduces the area for gas exchange and makes sufferers short of breath. People who work in dusty factories, mills, quarries or mines without protection, and people who smoke tobacco, are likely to get emphysema and other diseases of the breathing system.

### Other ways of breathing

- Mammals, birds and reptiles have lungs and breathe as we do.

- Amphibians have lungs but also exchange gases through their skin.

- Insects have tiny tubes that carry air into their bodies.

- Fish take in water through their mouths. The water passes over their **gills**. Oxygen dissolved in the water diffuses into the fish's blood. Carbon dioxide diffuses out of the blood and into the water, which then passes out through the gill slits.

### ◄ CHECKPOINT ►

1 Why do people with emphysema have to breathe pure oxygen?

_____

2 Why do fish have to keep water moving over their gills?

_____

## Respiration (8B1)

**1** Unscramble the letters to find the words needed to complete these sentences:

**biocare    cluesgo    yrgene**

    **a** Respiration is a chemical reaction that gives a cell _____

    **b** Respiration involving oxygen is said to be _____

    **c** In respiration oxygen reacts with _____

**2** Complete the word equation:

glucose + _____ → _____ + water + _____

**3** Explain why the following need to respire more than normal:

    **a** Pregnant women:

    _____

    **b** The root hair cells of a plant when fertiliser is added to the soil:

    _____

    **c** An Antarctic explorer:

    _____

    **d** A nurse lifting a patient into bed:

    _____

## The heart and circulation (8B2)

**1**   **a** Label the diagram of the heart with:

    **pulmonary vein (PV)    pulmonary artery (PA)**
    **left atrium (LA)    right atrium (RA)    left ventricle (LV)**
    **right ventricle (RV)    valve (V)    aorta (A)**

    **b** Draw arrows on the diagram to show the movement of
    the blood through the heart.

**2** Choose the answer to each of the following questions from:

arteries      veins      capillaries

    **a** Which blood vessels have thick walls? _____

    **b** Which blood vessels pass near every cell? _____

    **c** Which blood vessels carry blood to the heart? _____

    **d** In which blood vessels can you feel a pulse? _____

**3** What happens to your pulse rate when you exercise? Explain the change.

_____

## Supplying cells (8B3)

**1** Unscramble the letters to find the words that complete these sentences:

I laid c..c..cat    if fun is do    hay boxing mole
hale big moon    a nice boar

   **a** The process in which molecules move from a place of high concentration to a place of low

      concentration is called _____

   **b** The red substance in red blood cells is called _____

   **c** When red blood cells absorb oxygen it forms a substance called _____

   **d** Someone using up energy fast has to use _____ respiration, which produces

      _____ in the muscles.

**2** Sketch a diagram of a capillary, a red blood cell and a tissue cell (in the box). Show and name the substances moving each way across the capillary.

**3** What is fermentation?

_____

_____

_____

## Breathing (8B4)

**1**  **a** Cross out the incorrect word in each pair in the following paragraph:

     When you inhale, the muscles in your chest **contract/relax**. This makes your ribs **rise/fall** and your diaphragm becomes **domed/flatter**. The volume of the chest **decreases/increases**, so that air is pushed **into/out of** your lungs.

   **b** Write what happens when you exhale:

     _____

     _____

     _____

     _____

     _____

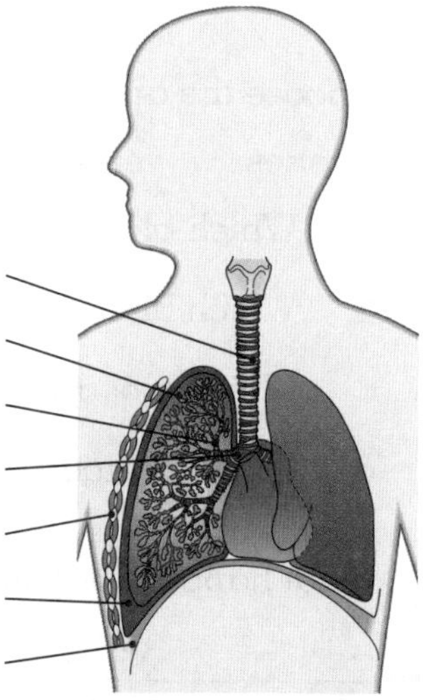

**2** The diagram shows the chest and breathing system.

Label the diagram with the following words:

**lung    trachea    bronchus    bronchiole
alveoli    ribs    diaphragm**

**1** What is the percentage of oxygen in:

a   inhaled air? _____

b   exhaled air? _____

**2** Which gas controls our rate of breathing?

_____

**3** Why do alveoli have:

a   thin walls?

_____

b   a large surface area?

_____

**4** The diagram shows an alveolus. Write on the diagram the gases that are exchanged in the lungs and the directions they move.

## Lung damage (8B6)

**Keyword clues**

1   _____ is the mixture of gases we breathe.

2   When you inhale you _____ in.

3   _____ fibrosis is an inherited disease that affects the breathing system.

4   The heart _____ blood into the blood vessels.

5   The disease in 3 makes _____ mucus that clogs up airways.

6   Asthma is a disease that makes the airways become

_____

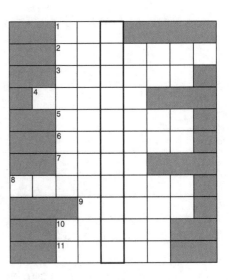

7 and 11   Dust from quarries and _____ (11)

damage alveoli and make _____ (7) tissue.

8   Fish can breathe because oxygen makes a _____ in water.

9   Fish take in oxygen when water passes over their _____

10   _____ from cigarettes causes breathing diseases.

**Keyword:** What this unit is all about.

1  a) The diagram shows the heart, lungs and blood vessels:

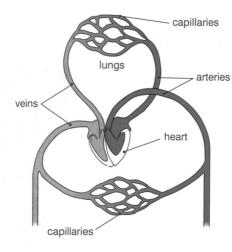

**i)** Why is the heart made of muscle?

> **HINT**
> Think of the job the heart has to do.

_____

_____  (1)

**ii)** Draw arrows on the diagram to show the direction the blood travels from the heart to the lungs. _____ (1)

**iii)** Which blood vessels have thick walls? _____ (1)

b) The diagram shows a capillary carrying blood near cells in tissue:

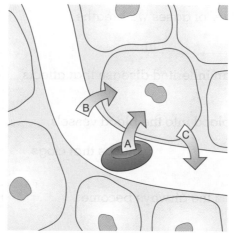

Substance A is carried by red blood cells.
Substance A and B leave the blood and enter cells.
Substance C leaves cells and enters the blood.
What are gases A, B and C?

> **HINT**
> Think about the substances needed in respiration.

**Gas A:** _____

**Gas B:** _____

**Gas C:** _____ (3)

**2** The diagram shows the exchange of gases that takes place in the lungs:

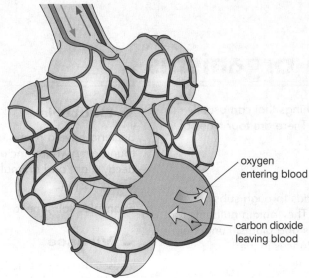

oxygen entering blood

carbon dioxide leaving blood

a) Which of the following is the name of the small sacks where gas exchange takes place?
Underline the correct name:

**trachea      bronchiole      alveolus      larynx**                                      (1)

b) Why do these sacks need to have a large surface area?

_____

_____                                      (1)

c) How can dust affect the breathing system?

_____

_____        (2)

HINT

Dust is hard, sharp particles of stone.

**3** When athletes start running their muscles need energy to move:

a) Which digested food provides the energy that muscle cells need?

_____        (1)

b) Why do athletes start to breathe faster as they start to run?

_____        (1)

c) After exercising hard, the muscles start to ache and may go into cramp. Even when the exercises end, breathing and heart rate stay high for a time.
Explain these observations:

HINT

What happens if a cell cannot get the oxygen it needs fast enough?

_____

_____

_____

_____        (3)

# Microbes and disease

## 8C1 Micro-organisms

Micro-organisms are tiny living things that can usually only be seen with a microscope. There are four types of micro-organism:

### Fungi

Fungi often form fine, white threads through substances and are made up of many cells. They obtain nutrients for growth from dead wood, animals and plants. Fungi help things to decay and recycle substances in the environment. They reproduce by making spores.

**Examples**: mushrooms, yeast.

### Protozoa

Protozoa are single-celled animals found in ponds and damp places. They feed on bacteria and small particles.

**Example**: amoeba.

### Bacteria

Bacteria have a round or capsule-shaped cell. They are found everywhere, including inside animals. Soil bacteria decay plant and animal remains. They reproduce by binary fission.

**Example**: salmonella.

### Viruses

Viruses are much smaller than bacteria and have to invade other cells to grow and reproduce. They are just a coating over a few genes. Viruses grow inside every other living thing.

**Examples**: cold virus, HIV.

> **◀ CHECKPOINT ▶**
>
> Make a list of the similarities and differences between the four types of micro-organism:
>
> _____
>
> _____
>
> _____
>
> _____

## 8C2 Growing micro-organisms

Fungi and bacteria will grow on almost anything, but to grow particular types of micro-organism special methods are needed:

- First **sterilise** all the apparatus to kill any micro-organisms that may be on it. This is done by heating to a high temperature for several minutes.

- Prepare a broth of nutrients. This can be liquid or a jelly made with **agar**. In the laboratory, the nutrient is put in shallow containers called **Petri dishes**. In industry, large tanks are used.

- Spread a sample of the bacteria or fungi over the surface of the jelly. This is called **inoculation**.

- Cover the dish and put it in an **incubator** to keep it warm for a few days.

- Look at the Petri dish. You should see clumps or **colonies** of bacteria. The number and size of the colonies gives an idea of the number of bacteria in the sample.

> **◀ CHECKPOINT ▶**
>
> 1 Why must the apparatus be sterilised?
>
> _____
>
> 2 What is nutrient agar for?
>
> _____
>
> 3 What does 'inoculation' mean?
>
> _____

# 8C3 Microbe factories

Micro-organisms can be useful. Particular types of microbe are encouraged to grow to make all sorts of products:

## Enzymes

All living cells produce enzymes. Bacteria can be encouraged to produce particular enzymes for use in washing powders, or to turn waste carbohydrates into sugar for drinks.

## Antibiotics

Bacteria and fungi produce drugs that can attack disease bacteria.

## Food

Bacteria and fungi help to make cheese and yoghurt.

## Yeast

Yeast is a very useful single-celled fungus. It gets energy from sugars by a form of **anaerobic respiration** called **fermentation**:

$$\text{sugar} \xrightarrow{\text{(yeast)}} \text{carbon dioxide} + \text{ethanol} \; (+ \text{ energy})$$

Yeast is used in bread making. The yeast feeds on the sugar in the dough, and gives off carbon dioxide gas that makes the dough 'rise'. Yeast is also used in making beer and wine, converting sugars to ethanol – the 'alcohol' in drinks.

**◄ CHECKPOINT ►**

Make a list of the products made using microbes that you use in one day.

_____

# 8C4 Harmful micro-organisms

Some micro-organisms cause diseases when they enter the body. These micro-organisms are called **pathogens.** Some pathogens and the diseases they cause are shown below:

| Type of micro-organism | Diseases |
|---|---|
| bacteria | food poisoning, TB, tetanus |
| virus | measles, chicken-pox, food poisoning, colds, flu |
| fungus | ringworm, athlete's foot |
| protozoa | food poisoning, malaria |

**◄ CHECKPOINT ►**

Describe how an infected person could pass on a pathogen.

_____

_____

When we pick up a pathogen from somewhere, we say we have been **infected.**

There are various ways that micro-organisms can travel from person to person. This is called **transmission.** Sewage is a source of pathogens.

| Method of transmission | Examples |
|---|---|
| **droplets** of water in the air sneezed or coughed by infected people | colds, flu, TB |
| **oral:** eating food that has been contaminated with pathogens | food poisoning (e.g. salmonella, listeriosis) |
| drinking **water** contaminated with pathogens | cholera, dysentery, typhoid |
| **touching** objects or people | impetigo, ringworm |
| **insect bites** | malaria |
| **wounds** | tetanus, rabies |

# 8C5 Defending your body

Our bodies have a few lines of defence against infection by pathogens:
- **Skin** is a barrier to pathogens. The outer layer of dry, dead skin cells are constantly flaking off, taking microbes with them.
- Eyes, mouth, nose, sweat glands and other openings allow microbes to enter but fluids, such as tears, contain **chemicals** that kill microbes.
- Once micro-organisms get inside the body, they are attacked by **white blood cells.** They recognise the chemicals on the body's own cells, and destroy anything that has different chemicals called **antigens**.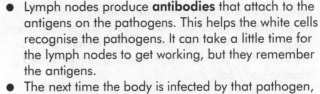

- Lymph nodes produce **antibodies** that attach to the antigens on the pathogens. This helps the white cells recognise the pathogens. It can take a little time for the lymph nodes to get working, but they remember the antigens.
- The next time the body is infected by that pathogen, the antibodies are produced quickly and the micro-organisms are destroyed before they can multiply. The body has become **immune** to the pathogen.

**◄ CHECKPOINT ►**

Draw a flow chart showing the defences that an invading pathogen must fight in order to infect a person.

# 8C6 Immunisation

**Immunisation** is a way of giving your body a head start in producing antibodies to fight infections. A **vaccine** makes you **immune** to the pathogen.

Children are vaccinated against tetanus, polio, diphtheria, measles, mumps and rubella.

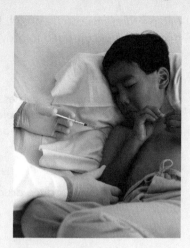

## Vaccines

Vaccines are made from dead micro-organisms, or bits of the microbe. Sometimes a weakened form of a virus can be used. The vaccine cannot cause the disease, but it contains **antigens** from the micro-organism that allow the lymph nodes to prepare antibodies. The antibodies are ready to fight the pathogen if the person did become infected.

People can be **inoculated** with the vaccine by an injection or by swallowing it.

Many diseases in the past, particularly those caused by viruses, had no cures and resulted in a lot of deaths. Immunisation prevented the microbes from multiplying, and the number of cases of the diseases has fallen. For example, the smallpox virus has been eliminated and polio is now very rare. Measles is the next target.

> **◄ CHECKPOINT ►**
>
> Describe how vaccination prevents disease.
>
> _____
>
> _____

# 8C7 Preventing infections

Vaccination can prevent some diseases, but there are plenty of other pathogens waiting to infect us.

- **Antibiotics,** such as penicillin, can help our bodies fight bacterial diseases, but they have no effect on viruses. Antibiotics must be taken until all the bacteria are destroyed to prevent them re-infecting.

- The best way to stop getting diseases is to prevent infection, that is, kill the micro-organisms before they get into your body. **Disinfectants** are powerful chemicals that damage micro-organisms and kill them. Disinfectants are used on surfaces where pathogens may be found, e.g. in kitchens and bathrooms. Common disinfectants are solutions of chlorine or ammonia.

- Disinfectants also harm human cells, so we use less powerful substances called **antiseptics** to deal with micro-organisms on our bodies. Antiseptics slow down the growth of microbes. Deodorants contain an antiseptic that stops microbes feeding on sweat and making body odour.

> **◄ CHECKPOINT ►**
>
> A small paper disc is soaked in the anti-microbial substance (disinfectant or antiseptic) and placed in a Petri dish of agar jelly, which has been inoculated with bacteria. The dish is incubated and examined after a few days.
>
>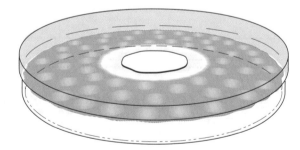
>
> How could you tell how effective a disinfectant or an antiseptic it is?
>
> _____
>
> _____

## Micro-organisms (8C1)

**1** Unscramble these words to find the micro-organism:

i b e a r c a t       s i r v u       a z o o p o r t       s u n f u g

_____   _____   _____   _____

**2** Look at the picture:

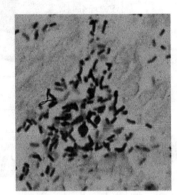

  **a** What type of micro-organisms are these?

    _____

  **b** Where might you find them?

    _____

  **c** How do they get energy?

    _____

  **d** How do they reproduce?

    _____

**3** Choose the micro-organism that matches these statements:

  **a** Must invade living cells to grow. _____

  **b** Single-celled predators that eat bacteria. _____

  **c** Are simply a few genes and a chemical coat. _____

## Growing micro-organisms (8C2)

**1** Allowing bacteria or fungus to grow on food could be dangerous.
Explain why this is.

_____

_____

**2** Explain the words:

**sterilise:** _____

**incubate:** _____

**inoculate:** _____

**3** Draw arrows linking the boxes in the correct order:

| Pour nutrient agar into Petri dish and allow it to set. |
| Examine the Petri dish for colonies of bacteria. |
| Inoculate agar with bacteria sample. |
| Sterilise the apparatus. |
| Incubate the bacteria for a few days. |

**4** Sketch the appearance of this Petri dish after colonies of bacteria have grown:

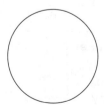

## Microbe factories (8C3)

**1** All the household products shown here are made using microbes. In each box, state what type of microbe is used and the job that it does:

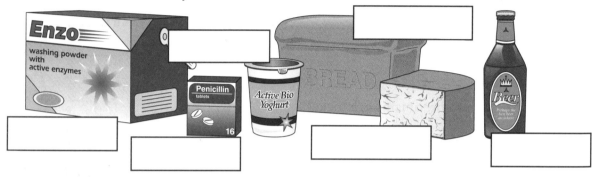

## Harmful micro-organisms (8C4)

**1** Explain the meaning of the words:

pathogen: _____

infection: _____

transmission: _____

**2** Unscramble the letters to find the disease and state which type of micro-organism causes it:

|  | t e a n u t s | a i r l a m a | s e a s e l m | g r o w n r i m |
|---|---|---|---|---|
| Disease: | _____ | _____ | _____ | _____ |
| Microbe: | _____ | _____ | _____ | _____ |

**3** How are the following diseases transmitted? (The picture gives a clue to one answer.)

colds: _____

malaria: _____

cholera: _____

salmonella poisoning: _____

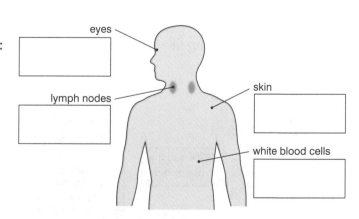

## Defending your body (8C5)

**1** Explain how each of the body's defences work against infection by micro-organisms:

eyes

lymph nodes

skin

white blood cells

**Words to use:**
pathogen
antigen
antibody
immunity

## Immunisation (8C6)

**1** What is the correct order for these statements?

   A:   Vaccine injected.

   B:   Body infected by live microbes.

   C:   Antibodies produced.

   D:   Vaccine made from dead microbes.

   E:   White blood cells destroy microbes.

   F:   Lymph nodes develop antibodies to vaccine antigens.

> **Correct order:**
> _____

**2** No one on Earth has suffered from smallpox for years. Why is this?

_____

**3** The number of children immunised against measles, mumps and rubella (MMR) dropped after the year 2000. What effect might this have on the number of cases of measles? Explain your answer.

_____

_____

_____

## Preventing infections (8C7)

**1** Explain the terms:

   **antibiotic:** _____

   **disinfectant:** _____

   **antiseptic:** _____

**2** a   Why aren't antibiotics given to people with flu?

   _____

   b   Why is it important to take antibiotics until an infection has completely disappeared?

   _____

**3** 'Sweet Pits' is a new underarm deodorant that claims to be an antiseptic. Describe how you could test this claim. You can draw a diagram in the box.

_____

_____

_____

_____

# APPLY YOUR KNOWLEDGE

**(SAT-style questions)**

**1** Mike wanted to make some bread. He mixed flour, sugar and yeast with a little water to make dough. He left the dough in a warm place for an hour. When he came back the dough had grown in size.

**a)** Yeast is a micro-organism. Underline the group of micro-organisms that yeast belongs to:

**bacteria** **fungi** **protozoa** **viruses** (1)

**b)** Why did Mike add some sugar to his dough?

(1)

**HINT** Think about how yeast gets energy and what is produced.

**c)** Explain what happened to make the dough rise.

(2)

**d)** Why was it a good idea for Mike to put the dough in a warm place?

(1)

**2** One evening Reese went to the cinema with her friends. The person sitting next to her sneezed a lot during the film. Next day Reese began to get a headache and started sneezing. She had been infected by the cold virus. A couple of days later she felt much better.

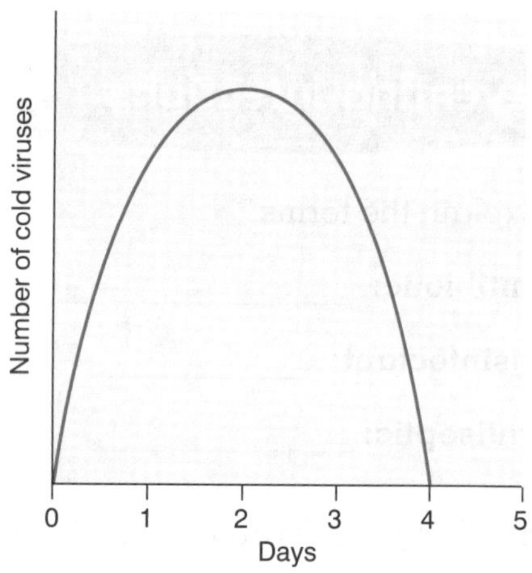

**a)** How did Reese catch the cold?

(1)

**b)** Use the graph to explain why Reese didn't feel unwell until the next day.

**HINT** What happens to the number of viruses in Day 1?

(1)

**c)** Reese's friend suggests that she goes to the doctor to get an antibiotic. Why won't this help Reese's cold?

**HINT** Antibiotics kill bacteria.

(1)

**d)** Explain why Reese got better.

(2)

**3** In the 1880s, Carlos Finlay discovered that yellow fever was caused by a virus carried by mosquitoes. The mosquitoes inject the virus when they bite the skin. The viruses quickly invade cells and produce poisons that make the victim ill.

**a)** Why do viruses invade cells?

_____ (1)

Finlay developed a vaccine. First he kept the viruses for a few days so that they became weakened. Then he injected the weakened viruses into some fit volunteers. Later, he got fresh mosquitoes to bite the volunteers. Most of the volunteers did not catch yellow fever.

**b)** Why did Finlay use weakened viruses for his vaccine?

_____ (1)

**c)** How does vaccination make someone immune to yellow fever?

_____

_____

_____ (2)

> **HINT**
> Someone who is immune cannot catch the disease.

Modern vaccines for diseases caused by viruses do not use the whole virus, but just bits of the coating around the virus genes.

**d)** Suggest **one** advantage of this kind of vaccine.

_____ (1)

**4** Pete and Molly wanted to test three brands of disinfectant. They prepared three Petri dishes with nutrient agar. Then Molly wiped Pete's hand with a piece of cotton wool and wiped the cotton wool over the agar in each Petri dish. They dipped a small paper disk in one of the disinfectants and placed the disk in the middle of one of the dishes. They did the same with the other disinfectants and then put the three dishes in an incubator for a few days.

Disinfo    ANTIBAC    Microgone

**a)** What is the nutrient agar for?

_____

_____ (1)

**b)** Why did Molly wipe Pete's hand?

_____

_____ (1)

**c)** What is a disinfectant?

_____

_____ (1)

**d)** How did Pete and Molly expect to compare the strengths of the three disinfectants?

_____

_____

_____ (2)

> **HINT**
> The stronger the disinfectant the more bacteria it can kill.

## 8D1 Identifying animals and plants

In any habitat, you are likely to find many different species of plants and animals. Many of the animals will be **invertebrates,** of which the **arthropods, molluscs** and **segmented worms** will be most common.

### ● Animals

- Many animals are **herbivores**.
- Many are **decomposers** that feed on dead plant and animal material. They break up leaves and recycle nutrients.
- There are also **predators** that prey on the decomposers.

### ● Plants

Most plants are **flowering plants**, which are divided into two groups:

- **Grassy**
- **Broadleaved**.

### ● Arthropods

Arthropods have segmented bodies, a hard outer skeleton, jointed legs; many change their shape as they grow.

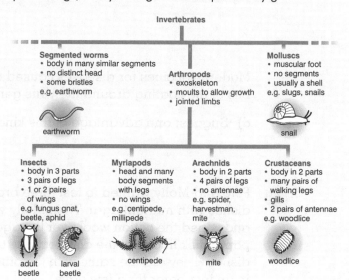

Invertebrates

**Segmented worms**
- body in many similar segments
- no distinct head
- some bristles
e.g. earthworm

earthworm

**Arthropods**
- exoskeleton
- moults to allow growth
- jointed limbs

**Molluscs**
- muscular foot
- no segments
- usually a shell
e.g. slugs, snails

snail

**Insects**
- body in 3 parts
- 3 pairs of legs
- 1 or 2 pairs of wings
e.g. fungus gnat, beetle, aphid

adult beetle    larval beetle

**Myriapods**
- head and many body segments with legs
- no wings
e.g. centipede, millipede

centipede

**Arachnids**
- body in 2 parts
- 4 pairs of legs
- no antennae
e.g. spider, harvestman, mite

mite

**Crustaceans**
- body in 2 parts
- many pairs of walking legs
- gills
- 2 pairs of antennae
e.g. woodlice

woodlice

#### ◄ CHECKPOINT ►

Check the meanings of:

Invertebrates: _____

Herbivores: _____

Predators: _____

Decomposers: _____

Arthropods: _____

Molluscs: _____

Flowering plants: _____

## 8D2 Sampling animals

Finding out about the **populations** of animals that live in a habitat needs a bit of ingenuity, since they move about:

- Large animals can be observed, recorded on camera, or caught in traps that do not harm them.

- Small animals can be caught in a **pitfall trap**, or shaken out of trees and bushes and collected in nets.

- To find the number and variety of small animals living on the ground we choose a small area, say 30 cm by 30 cm. The top 10 cm of leaf litter and soil can be removed and taken to the laboratory for investigation. Small brushes and pooters are used to move animals.

- When animals have been identified, large animals can be weighed and measured and small animals counted to find their **abundance**.

#### ◄ CHECKPOINT ►

Write out a plan for investigating a habitat.

_____

_____

_____

_____

Make a list of all the equipment you will need, and how you will record the population of the habitat.

# 8D3 Sampling plants

To find the **distribution** of plants in an area, we could record every plant and where it is growing. It is much simpler to take a **sample**. The sample can be from a small area, chosen at **random**.

A random sample area can be chosen by throwing an object over a shoulder and seeing where it lands, or by dividing up the large area into a grid and choosing a point in the grid using numbers such as birthdays or telephone numbers.

The points that are picked are the centre of the area to be investigated.

A **quadrat** is a square frame, usually 0.5 m by 0.5 m. They are placed over the random point.

All the plants within the quadrat are identified, counted and the area they cover **estimated**. The **percentage cover** can be worked out. The conditions (temperature, light intensity, moisture) are measured for each quadrat.

A **transect** is a long straight line across an area, and may cover a number of different habitats. Samples can be taken using quadrats at regular points along the line.

## ⟪ CHECKPOINT ⟫

1 Why are random samples taken?

2 Why should quadrats all have the same area?

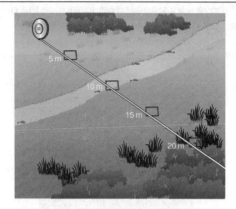

3 What does sampling a transect show?

# 8D4 Food webs

Animals usually eat more than one type of food. **Food webs** show the relationships between animals and plants in a habitat:

- At the base are the **producers**, which are plants.
- Above are **levels** of **consumers**.
- First are the **primary** consumers, i.e. the **herbivores** that eat plants.
- Above them are the **carnivores** that are **predators**.

The animals on each level **compete** with each other for sources of food.

**Decomposers** feed at all levels because they eat the dead remains and waste of animals, but they are mainly primary consumers because their main food is bits of dead plant.

## ● Changes in the web

Changes to a habitat can alter the relationships in a food web and increase competition. Changes in farming methods mean that hedgerows are dug up, or cut down, and fields extended. With fewer flowers providing nectar for insects, there will be less food for birds. Species that depend on one type of plant may disappear completely.

## ● Alien species

'Alien species' are plants and animals introduced into a habitat from other places, sometimes by gardeners, as pets, or by international trade. Alien species compete with native species. Alien plants may not provide food for native animals. Native species often become endangered.

**Threatened by 'alien' plants and animals?**

## ⟪ CHECKPOINT ⟫

Sketch the levels of a simple food web, showing where decomposers feed.

# 8D5 Interactions

Plants in a habitat compete with each other for light, space and minerals. Plants can **adapt** themselves to the conditions to help them survive. For example, if they are in the shade they may grow taller to reach the light, or grow larger leaves to take in more light.

Many plants are **specialised** for particular environments.

- Cacti have very small leaves to reduce the amount of water lost and store water in their stems.

- Marsh plants have roots adapted to grow in acidic, waterlogged soil where there is little oxygen.
- Nettles grow where there are plenty of nitrates – a vital mineral – from old dung heaps.

## Predators and prey

The populations of predators and prey in a habitat are closely linked. If there are few prey animals, then the number of predators will drop. Or, if the number of predators increases in a **population boom** the population of the prey animals will fall.

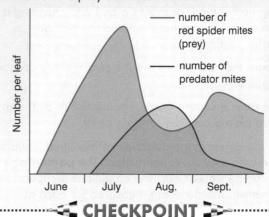

--- CHECKPOINT ---

What factors affect the populations of plant and animal species?

_____

_____

# 8D6 Pyramids and energy flow

A food web shows what animals eat, but ecologists need **quantitative data**. A **pyramid of numbers** shows the number of organisms at each level of a food web. Usually the bottom level is the broadest with the greatest number, however a single tree can provide food for a large number of consumers. The top level is usually very narrow, showing that there are few top predators. A top predator may have many parasites living on it.

(a) conventional          (b) inverted

top carnivores
intermediate carnivores
herbivores
producers

(c) including parasites

parasites
top carnivores
intermediate carnivores
herbivores
producers

## Energy transfer

Plants take in energy from the Sun to make glucose and starch. Some energy is used by respiration. Only a fraction of the energy taken in is passed along a food chain from producer to consumers. Animals use up

energy in movement, and some is lost in wastes such as urine. About 10% of the energy taken in by an animal is passed on to the next consumer. Less and less energy is passed up the food chain. Dead matter from plants and animals, and wastes, provide the energy for decomposers.

--- CHECKPOINT ---

How does energy transfer explain the pyramid of numbers?

_____

_____

## Identifying animals and plants (8D1)

**1** Identify the groups to which these animals belong:

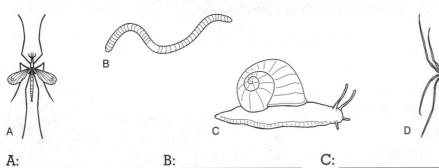

A: _____   B: _____   C: _____

D: _____   E: _____

**2** What features are shared by all arthropods?

_____

**3** Why isn't the Earth covered with dead leaves and the bodies of dead animals?

_____

## Sampling animals (8D2)

**1** Mike and Reese have chosen to investigate an area of woodland. They have found a badger set and would like to observe the badgers when they are active overnight. Describe how Mike and Reese could record the behaviour of the badgers.

_____

_____

**2** Mike and Reese have chosen a site in the woodland to do a survey of small animals:

   **a** What equipment do Mike and Reese need to collect animals from the trees?

_____

   **b** What equipment do Mike and Reese need for collecting animals on the ground?

_____

   **c** They took a sample of the leaf litter from the woodland floor back to the laboratory. How would Mike and Reese sort and measure the abundance of the animals in their sample? Mention the equipment that would be needed.

_____

_____

_____

_____

## Sampling plants (8D3)

**1**  Pip and Molly chose a random spot in a meadow to sample the plants. They placed a quadrat on their spot and Pip sketched what they saw in the quadrat.

**a**  Why did they choose a random spot for their investigation?

_____

**b**  What is a 'quadrat'?

_____

**c**  Look at Pip's sketch of the quadrat:

    **i**  Complete the table:

| Plant species | Number of plants |
|---------------|------------------|
| dandelion     |                  |
| clover        |                  |
| buttercup     |                  |
| thistle       |                  |

    **ii**  What percentage of the quadrat is covered by dandelions?

    _____

**d**  Would one quadrat give reliable data for the abundance of plants on the meadow? Explain your answer.

_____

**e**  Pip and Molly decided to look at more quadrats on a transect. What is a 'transect'?

_____

## Food webs (8D4)

**1**  The diagram shows part of a food web:

**a**  From the food web, name:

    **i**  a herbivore: _____

    **ii**  a carnivore: _____

    **iii**  a producer: _____

    **iv**  a predator: _____

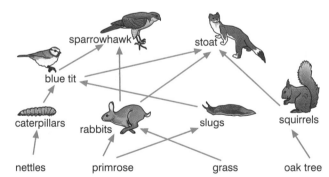

**b**  Mark on the diagram where decomposers, such as earthworms, fit in to the food web.

**c**  Describe the effect on the community if the nettles were cleared.

_____

**d**  What might be the effect of a disease that killed most of the rabbits?

_____

**e**  What could happen if an aggressive predator was accidentally released from a wildlife park?

_____

**1** Plants adapt to the conditions in which they grow. How would young pine trees respond when planted close together?

_____

**2** Plants are also specialised to cope with particular conditions. What conditions is a plant that grows on a sand dune beside the sea adapted to?

_____

**3** The graph shows how the numbers of a predator and prey varied over a few years:

a Mark on the graph a time when there was a population boom in the prey.

b Why does a population boom in predators follow a rise in prey numbers?

_____

_____

_____

c What happens to the numbers of prey after the boom in predators? Explain the pattern.

_____

_____

_____

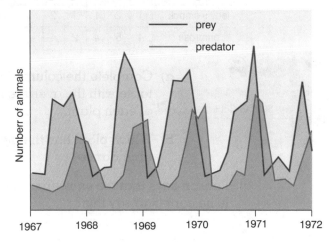

## Pyramids and energy flow (8D6)

**1** Look back at the food web in 8D4. Investigation showed that there were 45 caterpillars feeding on a patch of nettles. There were hundreds of nettle leaves in the patch. 8 blue tits and 1 sparrowhawk were seen in the same area.

a Sketch a pyramid of numbers in the box.

b What is the primary consumer in this food chain?

_____

c Why are there so few top predators?

_____

_____

**2** Fill in the gaps in this diagram of energy transfers in a food chain:

energy used for _____

energy from the

_____ → plants → _____ → carnivore → (after death) _____

dead leaves eaten by _____

# APPLY YOUR KNOWLEDGE

(SAT-style questions)

 Benson and Pip did a survey of the plants that grow beside a major road. They used a quadrat to sample small areas. They recorded the numbers of plants in each quadrat:

| Plant | Quadrat number | | | | | Average number |
|---|---|---|---|---|---|---|
| | 1 | 2 | 3 | 4 | 5 | |
| cowslips | 5 | 6 | 6 | 2 | 5 | |
| snowdrops | 3 | 2 | 3 | 0 | 2 | |
| lady's smock | 1 | 2 | 1 | 1 | 2 | |
| primrose | 4 | 5 | 4 | 2 | 4 | |

**HINT**

$$\text{average} = \frac{\text{sum of data}}{\text{}}$$

a) Complete the column in the table with the average number of each plant. (2)

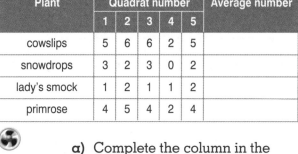

b) Which plant has the highest abundance? _____ (1)

c) Benson says that the snowdrops are evenly distributed over the roadside bank. Is he correct to say this?

_____ (1)

d) Which quadrat may have been taken nearest the road? Explain your answer.

_____ (2)

e) Benson notices that the plants on the bank on the opposite side of the road, which faces south, have smaller leaves. What reason can you give for this?

_____ (1)

**HINT**
Facing south means facing the Sun.

**HINT**
It can be dark amongst the grass.

f) The grass on the roadsides is not cut very often. How do the other plants adapt to this?

_____ (1)

2. Use the key to arthropods, shown below, to answer the following questions:

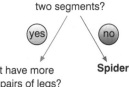

Does the animal have wings?
yes → **Insect**
no → Does it have more than two segments?
yes → Does it have more than 10 pairs of legs?
no → **Spider**
yes → **Myriapod**
no → **Crustacean**

a) Which group does not have wings and has just two segments?

_____ (1)

b) Centipedes are myriapods. Do centipedes have wings?

_____ (1)

c) A woodlouse does not fly, has many segments and ten pairs of legs. What group of arthropods do woodlice belong to?

_____ (1)

3　The diagram shows a food web in the South Atlantic Ocean:

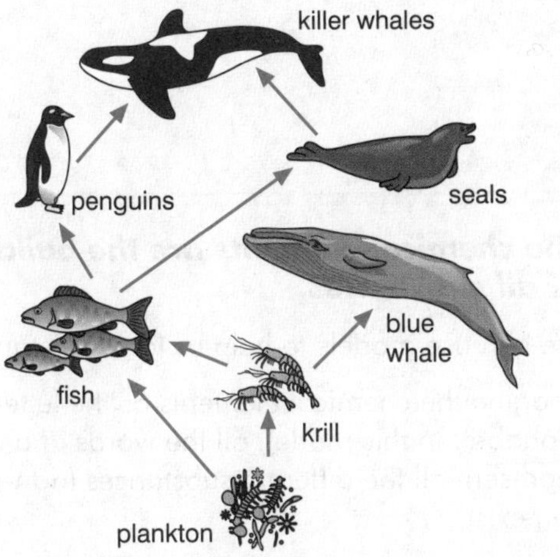

killer whales

penguins

seals

blue whale

fish

krill

plankton

**a)** Which organism is a producer?

_____ (1)

**b)** Underline the terms that could be applied to blue whales:

**herbivores　　consumers　　carnivores** (1)

**c)** What is the source of the plankton's energy?

_____

_____

_____ (1)

**d)** One food chain in the web is: plankton → krill → blue whale
Sketch a pyramid of numbers of this food chain. Use the box below. (2)

HINT

Start at the bottom.

HINT

Think about what the penguin does with the food that it eats.

**e)** Why are there very few killer whales compared to the number of penguins?

_____ (1)

**f)** Fishing trawlers are taking a lot of the fish out of the South Atlantic for us to eat. What effect could this have on the populations of animals in the South Atlantic?

_____

_____

_____ (3)

HINT

Look at all the organisms above and below the fish in the web.

**g)** What happens to bits of dead fish that the penguins and seals leave behind?

_____ (1)

**h)** The graph shows the population of penguins at an island in the South Atlantic.
Sketch a line to represent the population of killer whales on the same graph (the numbers do not have to be to scale). (1)

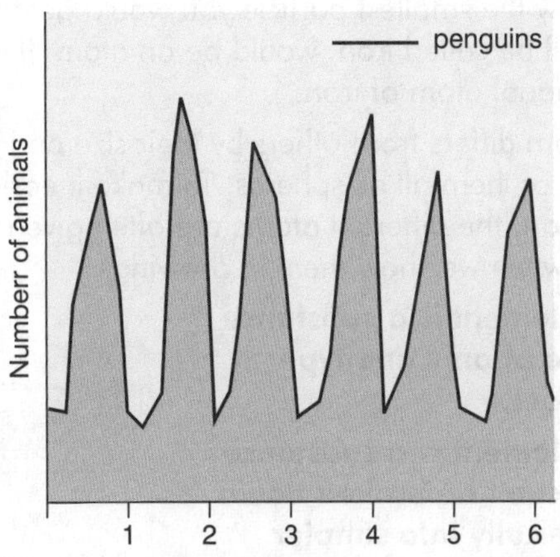

penguins

Numberr of animals

1　2　3　4　5　6
Years

# Atoms and elements

## 8E1 Different substances

### Nature's building blocks

Look at the substances below:

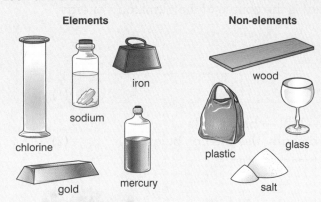

Elements | Non-elements

chlorine, sodium, iron, gold, mercury — wood, plastic, glass, salt

- There are some substances that we call **chemical elements** (or just elements). Examples are chlorine, iron, sodium, gold and mercury.

- Altogether, there are only 92 elements that occur naturally on Earth. About another 20 elements have been made artificially by scientists.

- All the other substances that exist (we can think of them as 'non-elements' for now) are made up from just these one hundred or so elements. The chemical elements are like nature's building blocks.

*The chemical elements are the building blocks of all substances.*

We can use models to help us to understand this.

Imagine the chemical elements as the letters of the alphabet. In this model, all the words in a dictionary represent all the different substances (non-elements) that can exist:

C + A + T

These letters represent three substances that are elements:

C A T

This word represents a substance that is one of the millions of non-elements.

### ◄ CHECKPOINT ►

Imagine that the chemical elements are the building bricks in a Lego set. Describe how this model explains the millions of substances that are non-elements.

_____

_____

_____

_____

## 8E2 Atoms

You've probably heard of 'atoms' before. The word 'atom' comes from a Greek word describing something that can't be divided up.

Imagine that you had a magic knife and started chopping up a piece of a chemical element, such as iron. You keep cutting and cutting until the bits of iron are really tiny – smaller than the smallest thing we can see through a normal microscope.

Eventually the smallest particle you would get to, that could still be called iron, would be an atom. It would be an individual atom of iron.

Each atom differs from others by their size and mass, but we think of them all as spheres. To make it easier to tell them apart, the different atoms are often given different colours when we show them in drawings.

- **An element is a substance made of only one type of atom.**

- **An element is a substance that can't be broken down chemically into simpler substances.**

### Chemical symbols

Notice the letter on the atom of sulphur. This is called the chemical **symbol** of the atom (or of the element). To a chemist, the symbol H represents one atom of the element hydrogen. Here are the symbols of some common atoms:

| Atom | Symbol | Atom | Symbol |
|---|---|---|---|
| hydrogen | H | zinc | Zn |
| carbon | C | iron | Fe |
| nitrogen | N | sodium | Na |
| sulphur | S | potassium | K |
| oxygen | O | copper | Cu |
| chlorine | Cl | helium | He |

**An atom of sulphur**

### ◄ CHECKPOINT ►

Give four symbols that are not based on the English name.

_____

# 8E3 Elements

## Sorting out the elements

In 1869 a Russian chemist called Dmitri Mendeleev arranged the elements in order of atomic mass. He started with the lightest atoms, getting heavier. He formed new rows so that similar elements lined up in vertical columns. The columns are called groups.

**Dmitri Mendeleev**

- He called this arrangement the **'Periodic Table'**.
- Periodic means 'repeated at regular intervals'.

Here is a modern version of his table:

- Most of the elements are **metals**. They are shaded in blue in the Periodic Table below.
- The rest are **non-metals**, with a few elements classified as **metalloid** or **semi-metals**.

Metalloids are on the borderline between metals and non-metals. They have some metallic and some non-metallic properties. Silicon is the most well known metalloid.

### ◄ CHECKPOINT ►

Sort these elements into three sets, labelling each set of elements:

iodine (I)   sodium (Na)   zinc (Zn)
aluminium (Al)   germanium (Ge)
nitrogen (N)   sulphur (S)   gold (Au)

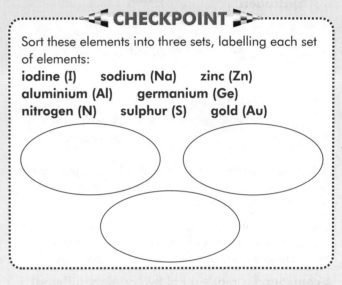

key:
- ☐ metals
- ☐ non-metals
- ☐ metalloids (semi–metals)

| | | | | | | | | | | | | | | | | | |
|---|---|---|---|---|---|---|---|---|---|---|---|---|---|---|---|---|---|
| H Hydrogen | | | | | | | | | | | | | | | | | He Helium |
| Li Lithium | Be Beryllium | | | | | | | | | | | B Boron | C Carbon | N Nitrogen | O Oxygen | F Fluorine | Ne Neon |
| Na Sodium | Mg Magnesium | | | | | | | | | | | Al Aluminium | Si Silicon | P Phosphorus | S Sulphur | Cl Chlorine | Ar Argon |
| K Potassium | Ca Calcium | Sc Scandium | Ti Titanium | V Vanadium | Cr Chromium | Mn Manganese | Fe Iron | Co Cobalt | Ni Nickel | Cu Copper | Zn Zinc | Ga Gallium | Ge Germanium | As Arsenic | Se Selenium | Br Bromine | Kr Krypton |
| Rb Rubidium | Sr Strontium | Y Yttrium | Zr Zirconium | Nb Niobium | Mo Molybdenum | Tc Technetium | Ru Ruthenium | Rh Rhodium | Pd Palladium | Ag Silver | Cd Cadmium | In Indium | Sn Tin | Sb Antimony | Te Tellurium | I Iodine | Xe Xenon |
| Cs Caesium | Ba Barium | La Lanthanum ► | Hf Hafnium | Ta Tantalum | W Tungsten | Re Rhenium | Os Osmium | Ir Iridium | Pt Platinum | Au Gold | Hg Mercury | Tl Thallium | Pb Lead | Bi Bismuth | Po Polonium | At Astatine | Rn Radon |
| Fr Francium | Ra Radium | Ac Actinium ►► | | | | | | | | | | | | | | | |

| | | | | | | | | | | | | | |
|---|---|---|---|---|---|---|---|---|---|---|---|---|---|
| Ce Cerium | Pr Praseodymium | Nd Neodymium | Pm Promethium | Sm Samarium | Eu Europium | Gd Gadolinium | Tb Terbium | Dy Dysprosium | Ho Holmium | Er Erbium | Tm Thulium | Yb Ytterbium | Lu Lutetium |
| Th Thorium | Pa Protactinium | U Uranium | Np Neptunium | Pu Plutonium | Am Americium | Cm Curium | Bk Berkelium | Cf Californium | Es Einsteinium | Fm Fermium | Md Mendelevium | No Nobelium | Lr Lawrencium |

**The Periodic Table of elements**

# 8E4 Combining atoms

- We call the links between atoms 'bonds'. When atoms join together, we say that the atoms **bond** to each other.

- When atoms bond together, the new particles they make are called **molecules**.

Look at the two molecules below:

**A molecule of hydrogen**

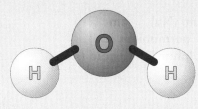

**A molecule of water**

Notice that the atoms in a molecule of hydrogen are both the same. We can say that hydrogen gas is made up of molecules of an element.

A water molecule contains two types of atom – hydrogen and oxygen. When a substance is made up of two or more different types of atom, we call it a **compound**.

**A compound is made up of two or more different types of atom.**

## Chemical formulae

Scientists use their own shorthand way of showing a molecule. Rather than drawing a molecule, they use its **chemical formula**. This shows us how many of each type of atom there are in a molecule. The formula does this by using the symbols of atoms, and subscript numbers.

For example, the formula of carbon dioxide is $CO_2$. Each molecule is made up of 1 carbon atom and 2 oxygen atoms.

**A molecule of carbon dioxide – its formula is $CO_2$**

If a molecule contains just one atom of a certain element, we don't bother writing a number 1 in its formula.

# 8E5 Reacting elements

## From elements to compounds

You have seen elements reacting together to make compounds. In Year 7, you saw different elements react with oxygen. The compounds made are called **oxides**.

- We can describe reactions by **word equations**. For example:

    magnesium + oxygen → magnesium oxide

- Remember that the substances we start with before the reaction are called **reactants**. The substances formed in reactions are called **products**.

Other non-metals, such as sulphur, chlorine and bromine, also react with other elements to form compounds.

- Sulphur makes compounds called **sulphides**; for example, magnesium sulphide.

- Chlorine makes **chlorides**.

- Bromine makes **bromides**.

# 8E3 Elements

## Sorting out the elements

In 1869 a Russian chemist called Dmitri Mendeleev arranged the elements in order of atomic mass. He started with the lightest atoms, getting heavier. He formed new rows so that similar elements lined up in vertical columns. The columns are called groups.

**Dmitri Mendeleev**

- He called this arrangement the **'Periodic Table'**.

- Periodic means 'repeated at regular intervals'.

Here is a modern version of his table:

- Most of the elements are **metals**. They are shaded in blue in the Periodic Table below.

- The rest are **non-metals**, with a few elements classified as **metalloid** or **semi-metals**.

Metalloids are on the borderline between metals and non-metals. They have some metallic and some non-metallic properties. Silicon is the most well known metalloid.

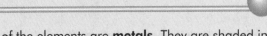

### ◄ CHECKPOINT ►

Sort these elements into three sets, labelling each set of elements:

iodine (I)     sodium (Na)     zinc (Zn)
aluminium (Al)     germanium (Ge)
nitrogen (N)     sulphur (S)     gold (Au)

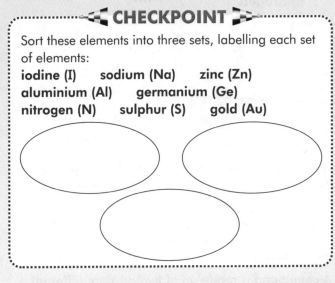

**The Periodic Table of elements**

# 8E4 Combining atoms

- We call the links between atoms 'bonds'. When atoms join together, we say that the atoms **bond** to each other.

- When atoms bond together, the new particles they make are called **molecules**.

Look at the two molecules below:

**A molecule of hydrogen**

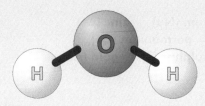 **A molecule of water**

Notice that the atoms in a molecule of hydrogen are both the same. We can say that hydrogen gas is made up of molecules of an element.

A water molecule contains two types of atom – hydrogen and oxygen. When a substance is made up of two or more different types of atom, we call it a **compound**.

**A compound is made up of two or more different types of atom.**

## Chemical formulae

Scientists use their own shorthand way of showing a molecule. Rather than drawing a molecule, they use its **chemical formula**. This shows us how many of each type of atom there are in a molecule. The formula does this by using the symbols of atoms, and subscript numbers.

For example, the formula of carbon dioxide is $CO_2$. Each molecule is made up of 1 carbon atom and 2 oxygen atoms.

**A molecule of carbon dioxide – its formula is $CO_2$**

If a molecule contains just one atom of a certain element, we don't bother writing a number 1 in its formula.

### ◄ CHECKPOINT ►

You've probably heard people talk about 'H-two-O' and know that it is water. We write this formula as $H_2O$. Explain why the formula of water is $H_2O$.

_____

_____

# 8E5 Reacting elements

## From elements to compounds

You have seen elements reacting together to make compounds. In Year 7, you saw different elements react with oxygen. The compounds made are called **oxides**.

- We can describe reactions by **word equations**. For example:

  magnesium + oxygen → magnesium oxide

- Remember that the substances we start with before the reaction are called **reactants**. The substances formed in reactions are called **products**.

### ◄ CHECKPOINT ►

Look at the word equation above and name:
a) the element that is a non-metal

_____

b) the compound

_____

c) the element that is a metal

_____

Other non-metals, such as sulphur, chlorine and bromine, also react with other elements to form compounds.

- Sulphur makes compounds called **sulphides**; for example, magnesium sulphide.

- Chlorine makes **chlorides**.

- Bromine makes **bromides**.

### ◄ CHECKPOINT ►

a) What do you think that we call the compounds made from iodine?

_____

b) Write a word equation that shows the reaction between magnesium and sulphur:

_____

## Different substances (8E1)

**1 a** Only 92 different elements exist naturally on Earth. So explain how we can have millions of different substances.

_____

_____

_____

**b** Use the letters 'T', 'R' and 'A' as elements to model your answer to part **a** above.

_____

_____

_____

**2** Sort these substances into two sets in the table opposite (don't forget to put the headings in each column):

**gold   salt   wood   chlorine   mercury   plastic   sodium**

| | |
|---|---|
| | |
| | |
| | |

## Atoms (8E2)

**1** Which of the statements could be used to describe what we mean by a 'chemical element'?

Circle the correct letters.

**A** An element is a substance made of only one type of atom.

**B** An element is a substance made from lots of different types of atom.

**C** An element is a substance that contains no atoms at all.

**D** An element is a substance that can't be broken down into simpler substances.

**2** Unscramble these anagrams to find the name of an atom and give its chemical symbol:

| Anagram | Atom | Symbol |
|---|---|---|
| N I C Z | _____ | _____ |
| E R G O N T I N | _____ | _____ |
| P R E P O C | _____ | _____ |
| R H U L U P S | _____ | _____ |
| N O I R | _____ | _____ |
| S T O P A I M U S | _____ | _____ |
| R O N B A C | _____ | _____ |
| M U D I O S | _____ | _____ |
| G N Y D H O R E | _____ | _____ |
| N E O G Y X | _____ | _____ |

1  Fill in the gaps in the following sentences:

It was in the year _____ that a chemist called _____ Mendeleev

arranged the chemical _____ in order of their atomic _____. He started

new _____ so that similar elements lined up in vertical _____

called groups.

The word 'periodic' means '_____ at regular intervals', so he called his

arrangement the **Periodic** _____.

2  Write the names of two substances in each
area of this Venn diagram:

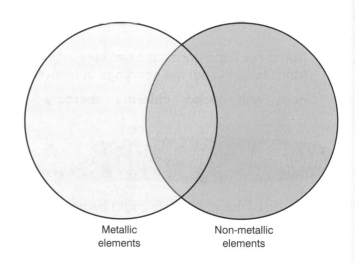

Metallic          Non-metallic
elements          elements

1  Define these words:

a  **Molecule:**

_____

_____

b  **Compound:**

_____

2  Draw a concept map, labelling the links
between these words:

| Atom |

| Element |          | Compound |

| Molecule |

## Combining atoms (8E4 CONTINUED)

**3** What is the formula of these molecules?

a

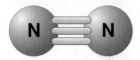

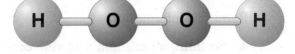

i The formula of nitrogen is ________

ii The formula of hydrogen peroxide is ________

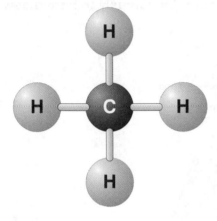

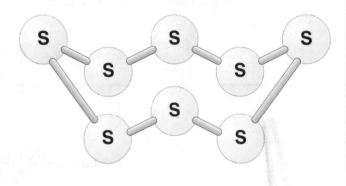

iii The formula of methane is ________

iv The formula of sulphur is ________

b What would be the best way to describe the particles that make up nitrogen and sulphur? Circle the correct letter:

**A** **Atoms of compounds**

**B** **Molecules of compounds**

**C** **Molecules of elements**

c What would be the best way to describe the particles that make up methane and also hydrogen peroxide? Circle the correct letter:

**A** **Atoms of compounds**

**B** **Molecules of compounds**

**C** **Molecules of elements**

## Reacting elements (8E5)

**1** Fill in the blanks below:

a ________ + oxygen → calcium oxide

b magnesium + ________ → magnesium chloride

c iron + bromine → ________ ________

d ________ + ________ → hydrogen chloride

e ________ + iodine → aluminium ________

**2** The answers to question 1 are all examples of ________ ________, which show ________ reacting with each other to form ________.

APPLY YOUR KNOWLEDGE

**1** Read this information:

- Sodium, aluminium and zinc are elements that conduct electricity.

- Iodine and sulphur are elements that do not conduct electricity.

- If zinc and sulphur are heated together, they react vigorously to form a new substance called zinc sulphide.

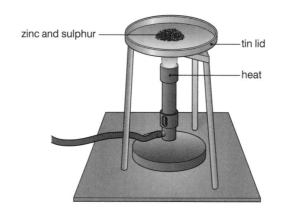

zinc and sulphur — tin lid

— heat

HINT

Do metallic or non-metallic elements conduct electricity?

a) Using the information given above:

   **i)** Name two metals:

      _____ and _____ (2)

   **ii)** Name two non-metals:

      _____ and _____ (2)

   **ii)** Give the name of a compound:

      _____ (1)

b) **i)** Write a word equation for the reaction of zinc with sulphur:

      _____ (1)

 HINT

Word equations always show:
reactants → products

   **ii)** Why would you carry out the reaction between zinc and sulphur in a fume-cupboard?

      _____

      _____ (2)

 HINT

Why might you need the glass screen of the fume-cupboard between you and the reacting elements? Why would you need the fan in the fume-cupboard? Think of what forms when sulphur is heated.

 HINT

Remember that sulphur changes the end of its name in its compounds – an example is given at the start of this question.

   **iii)** Write the name of the compound formed when magnesium reacts with sulphur.

      _____ (1)

**2** Look at this outline of the areas in the Periodic Table:

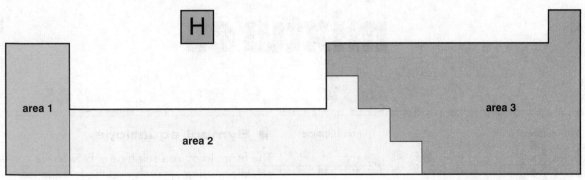

**a)** What does the symbol H stand for?

_____ (1)

**b)** In which areas of the Periodic Table would you find:

**i)** metallic elements?

_____ (2)

You don't need to know the symbols of all the elements in the Periodic Table (phew!) – but you should learn the common ones (such as those given in the table on p.34).

**ii)** non-metals, such as nitrogen and phosphorus?

_____ (1)

**iii)** very reactive metals, such as sodium and potassium?

_____ (1)

Which area shows the first two groups in the Periodic Table? This is where we find the most reactive metals.

**iv)** less reactive metals, such as iron and zinc.

_____ (1)

The less reactive metals are found between the reactive metals and the non-metals in the Periodic Table.

**c)** Why is sodium chloride not found in the Periodic Table?

_____

_____ (1)

You will find sodium and chlorine in the Periodic Table. These are both _____ whereas sodium chloride is a _____.

# Compounds and mixtures

## 8F1 Making compounds and mixtures

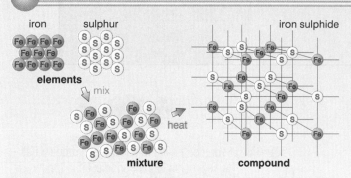

iron    sulphur                iron sulphide

**elements**

mix

**mixture**        **compound**

heat

- The ratio of iron and sulphur atoms in any sample of the compound iron sulphide will always be the same.

- The ratio in this case is 1:1 and that's why its formula is FeS.

- However, the proportions in a mixture can vary. We could mix one part iron with two parts sulphur, or three parts iron with one part sulphur, and so on.

*Any compound always has fixed proportions of its elements (as shown by its chemical formula).*

### ● Symbol equations

The formula of iron sulphide is FeS. So we can write an equation, using symbols and the formula, showing how we can make iron sulphide:

$$Fe + S \rightarrow FeS$$

- We call this a **symbol equation**.

- Notice that we have the same number of each type of atom on either side of the equation. We can then say that this is a 'balanced symbol equation'.

**CHECKPOINT**

Write a symbol equation for the reaction between zinc and sulphur:

_____

## 8F2 Reacting compounds

We also refer to reactions as 'chemical changes'. These are the opposite of 'physical changes'.

Here is a table showing the differences between chemical and physical changes:

| Chemical changes | Physical changes |
|---|---|
| new substances are formed | no new substances are formed |
| often cannot be reversed | usually easy to reverse |

**CHECKPOINT**

Look at a molecule of the compound we call water:

What would we get formed if we carried out a reaction to break down the compound?

_____

### ● Compounds and mixtures

Here is a table to summarise the differences between compounds and mixtures.

| Compounds | Mixtures |
|---|---|
| Have a fixed composition (will always have the same proportion of elements in any particular compound) | Have no fixed composition (the proportions vary depending on the amount of each substance mixed together) |
| Need chemical reactions to separate the elements in them | The substances can be separated again more easily (by physical means using the differences in properties of each substance in the mixture) |
| Are single substances | Contain two or more substances |
| Have properties different to those of the elements in the compound | Have properties similar to those of the substances in the mixture |

# 8F3 More about mixtures

## Raw materials

Here are some useful mixtures that we use as **raw materials**. These are the starting materials used in the chemical industry to make new products:

● Air

● Sea water

● Crude oil

● Rocks (some are called **ores**).

Air is not a single substance, but a mixture of different gases. Here are the gases we find in air:

| Gases in the air | Formula of gas | Approximate proportions |
|---|---|---|
| nitrogen | $N_2$ | 78% |
| oxygen | $O_2$ | 21% |
| carbon dioxide | $CO_2$ | about 0.04% |
| water vapour | $H_2O$ | (varies) |
| argon (about 0.9%) and other noble gases | Ar, He, Ne, Kr, Xe, Ra | 1% |
| various pollutants | e.g. $SO_2$ or $NO_2$ or $CH_4$ | |

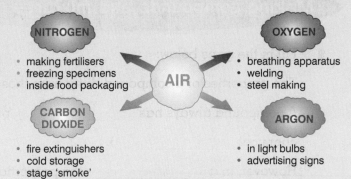

NITROGEN
● making fertilisers
● freezing specimens
● inside food packaging

OXYGEN
● breathing apparatus
● welding
● steel making

CARBON DIOXIDE
● fire extinguishers
● cold storage
● stage 'smoke'

ARGON
● in light bulbs
● advertising signs

## Separating the gases from liquid air

We can separate and collect liquids with different boiling points. The process is called **fractional distillation**. To do this to air we need to cool it down to condense it into a liquid. It's not easy, because the air has to be cooled to a temperature of almost −200 °C.

◄ CHECKPOINT ►

a) What is the main gas in air?

_____

b) Name the process we use to separate the gases in air?

_____

# 8F4 Pure or impure?

We can show the melting point and boiling point of a substance on a temperature line:

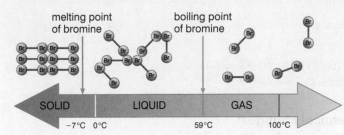

melting point of bromine

boiling point of bromine

| SOLID | LIQUID | GAS |

−7°C  0°C          59°C          100°C

This shows that the element bromine is a solid below −7 °C. It melts at −7 °C, turning into liquid bromine. The liquid boils at 59 °C, turning into a gas. So at room temperature (taken as 20 °C) bromine is a liquid.

## Pure substances and mixtures

● The melting and boiling points of an element or a compound are called its **fixed points**.

● Pure substances can be compounds or elements, but they contain only one substance. An **impure** substance is a mixture of two or more different elements or compounds.

● We can use melting points or boiling points to identify substances. That's because pure substances have characteristic temperatures at which they melt and boil.

● The melting point and boiling point of a mixture will vary depending on the composition of the mixture.

● A mixture does not have a sharp melting point or boiling point. It changes state over a range of temperatures.

I'm glad ice cream is a mixture... Imagine if it had a sharp melting point!

◄ CHECKPOINT ►

Which of these substances will have a sharp melting point? Why?

**chocolate    ice cream    salt    zinc**

_____ because _____

_____

### Making compounds and mixtures (8F1)

**1** Fill in the gaps below:

The properties of a compound differ from those of the _____ it is made from.

A compound always has _____ proportions of each element (shown in its

chemical _____).

However, in a _____ the proportions can _____.

**2** Draw the molecules in the boxes below:

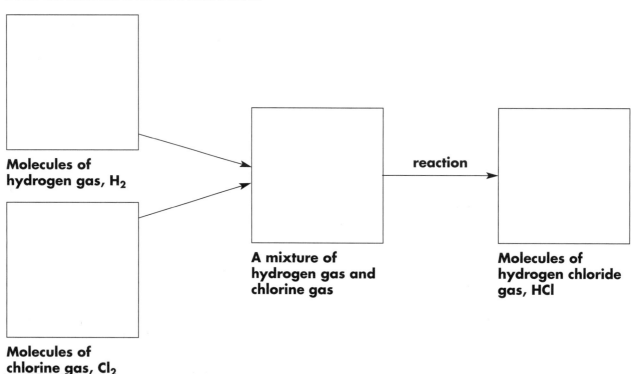

**Molecules of
hydrogen gas, H$_2$**

**Molecules of
chlorine gas, Cl$_2$**

**A mixture of
hydrogen gas and
chlorine gas**

reaction

**Molecules of
hydrogen chloride
gas, HCl**

### Reacting compounds (8F2)

**1** Label these changes as physical changes or chemical changes:

**a** burning a match: _____

**b** melting ice: _____

**c** alcohol evaporating: _____

**d** acid neutralising an alkali: _____

**2** List three differences between compounds and mixtures of elements:

_____

_____

_____

## More about mixtures (8F3)

**1** Draw a line between the raw material and one of its products:

| Raw material | Product |
|---|---|
| Crude oil | Salt (sodium chloride) |
| Rock (ore) | Iron metal |
| Air | Petrol |
| Sea water | Nitrogen |

**2** Circle the gas we use inside light bulbs:

oxygen          argon          nitrogen

carbon dioxide          hydrogen          helium

## Pure or impure? (8F4)

**1** Look at this table:

| Element | Melting point (°C) | Boiling point (°C) |
|---|---|---|
| chlorine | −101 | −34 |
| sodium | 98 | 900 |
| iron | 1540 | 2760 |

Which element or elements are:

**a** a gas at 20°C? _____

**b** a solid at 20°C? _____

**c** a liquid at −50°C? _____

**d** a liquid at 2000°C? _____

**e** a gas at 1000°C? _____

**2** The labels have fallen off the jars of two white powders. One is a mixture and one is a compound.

How could you decide which is which?

_____

_____

_____

**1** Look at the boxes below:

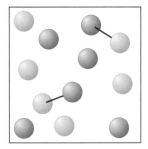

**Box A**

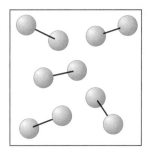

**Box B**

HINT

To answer this question (without just guessing!) you need to understand the words 'atom', 'molecule', 'element' 'mixture' and 'compound'.

**Box C**

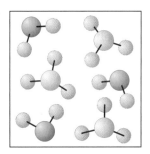

**Box D**

a) Which boxes contain mixtures?

_____ (2)

HINT

Remember that you can have pure compounds as well as pure elements.

b) Which boxes contain a pure substance?

_____ (2)

c) Which box contains a pure element?

_____ (1)

HINT

A reaction happens when atoms join together, or molecules break down, and atoms get rearranged.

d) Which box contains a mixture of compounds?

_____ (1)

e) Which box shows a chemical reaction taking place?

_____ (1)

f) Which box contains ammonia, NH₃?

_____ (1)

HINT

How many atoms are in an ammonia molecule? You can work it out because you are told that its formula is NH₃.

g) Which box contains some single atoms, not bonded?

_____ (1)

**2** The diagrams below show the different arrangement of atoms in six substances. Each atom is represented by a circle labelled with its chemical symbol.

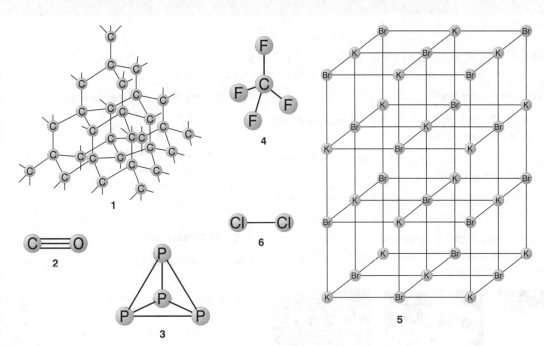

a) i) Which of the diagrams represent the structures of chemical elements?
Write down the correct numbers.

_____ (1)

ii) Explain how you decided which were elements.

_____

_____ (1)

HINT

What can you say about the atoms in all elements?

b) Give the formula of two of the compounds represented in the diagram.

_____ (2)

HINT

Remember that we don't put 1s in a chemical formula.

c) Give the name of the substance labelled 5.

_____ (1)

HINT

Don't forget that bromine changes the end of its name in its compounds.

d) Give the names of the chemical elements whose atoms can be represented by the following symbols:

i) C _____

ii) Cl _____

iii) Cu _____ (3)

# Rocks and weathering

## 8G1 Looking at rocks

### What are rocks?

There are many different rocks. They are usually made up of different **mixtures of minerals**.

- **A mineral is a solid compound or element found naturally in the ground.**

- For example, diamond is a mineral that is an element. It is an element made up of carbon atoms.

- On the other hand, halite is a mineral that is a compound. It is made up of sodium chloride. Most minerals are compounds.

**Diamond is a mineral that is an element. It is one form of the element carbon (C)**

**Halite is a mineral that is a compound. Its chemical name is sodium chloride (NaCl)**

**Granite is a rock made from a mixture of minerals**

### Texture

The 'texture' of a rock describes the way its grains fit together. There are two main types of texture in rocks:

- **Crystalline texture**: The mineral grains are crystals in the rock. The grains all interlock. There are no gaps between the crystals.

- **Fragmental texture**: The minerals form randomly shaped fragments or grains that do not fit together neatly. Another mineral often 'cements' the grains to each other.

### Porosity

- The **porosity** of a rock tells us about its ability to soak up water.

- Rocks that have spaces between their grains can soak up water better than rocks with interlocking crystals. The water fills the gaps between grains in rocks like sandstone.

**Porous rocks have grains that do not interlock**

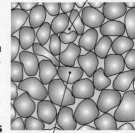

water in the gaps

grains in the rock

> ◄ **CHECKPOINT** ►
>
> Explain why sandstone is a porous rock.
>
> _____
>
> _____
>
> _____

## 8G2 Chemical weathering

- **Weathering** is the breakdown of a rock.

- Carbon dioxide gas dissolves slightly in water. It forms a **weakly acidic** solution of carbonic acid. Over time, this acid can chemically attack some of the minerals in rocks. As in all reactions, new substances are made.

- Limestones (which include chalk) and marble contain the mineral **calcite**. Its chemical name is calcium carbonate ($CaCO_3$).

- Other rocks also contain carbonates, such as magnesium carbonate or copper carbonate.

- Acids react with carbonates. They form a salt, plus carbon dioxide and water. The salt formed is often soluble in water.

So any carbonate in the rock breaks down in acid and forms a solution. That's how carbonate rock is weathered by acids in soil or rainwater.

Granite is a mixture of three types of mineral:

- Quartz

- Feldspar

- Mica.

The acid in rainwater attacks the feldspar and mica minerals. Eventually the granite is weathered into small particles of clay. These are carried away, along with any soluble compounds formed in solution, by the water.

> ◄ **CHECKPOINT** ►
>
> Fill in the gaps:
>
> When rocks are _____ down in nature,
>
> we say that the rocks have been _____ .

# 8G3 Physical weathering

## Forces that break down rock

We have seen how rocks can be broken down by chemical changes to their minerals.

They are also broken down in nature by forces caused by physical processes. We call this **physical weathering**.

## Freeze/thaw

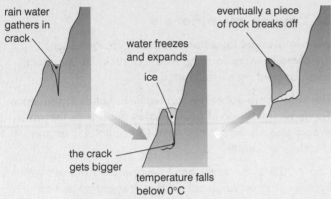

**The effect of freeze/thaw**

## Changing temperatures

- Rocks are mixtures of minerals. During heating and cooling, each of the minerals expands and contracts at different rates.

- This sets up stress forces within the rock. Eventually this causes the surface to crack and break away.

- On individual pieces of rock we can get an effect called **'onion-skinning'**. The rocks split into layers. They eventually fall away, like peeling off layers of an onion.

- Large masses of rock flaking off is called **exfoliation**.

It is important to realise that rocks will undergo many types of weathering at the same time.

**⫸ CHECKPOINT ⫷**

Name two types of physical weathering:

_____

# 8G4 Fragments in motion

## Transporting fragments

The fragments or minerals in solution, formed by weathering, get moved to another place. They are **transported** by:

- Gravity
- Wind
- Ice (in glaciers – rivers of ice)
- Water (in streams, rivers and seas).

## Depositing sediments

Weathered rock is often carried away in streams in rivers:

- At first, high in the mountains, fast flowing water races down steep slopes. Here it has enough energy to carry quite large pieces of rock. They bounce along the bottom of the river.

- The river starts to slow down as the land levels off. Then the larger bits of rock get deposited on the river bed. Deposited pieces of rock are called **sediments**.

- The smaller pieces of rock get carried along further before they are also deposited.

- So the **greater the energy** of the moving water, the **larger** the rock fragments it can carry along.

- The fine bits of rock, such as clay, can be carried all the way to the river mouth (estuary) where it meets the sea.

## Erosion

- **Erosion** is the wearing away of rock as surfaces rub against each other.

- Weathered fragments will erode the rock that they pass over. For example, rocks in river beds will be worn away as fragments scrape away at them.

- The weathered fragment itself also gets worn down (eroded). It will get smaller and smoother, the longer it gets carried along by the river.

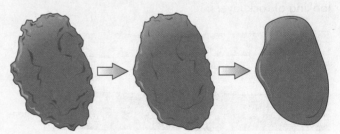

**⫸ CHECKPOINT ⫷**

List four ways that weathered bits of rock can be moved to other places:

_____

_____

What is a 'sediment'?

_____

# 8G5 Forming layers

- Different types of sediment form layers of different rocks.

Look at these two rocks taken from different layers in a cliff face:

**Sandstone rock**

**Conglomerate rock**

Look at the photo showing layers of rock formed from sediments deposited millions of years ago:

**Layers of sediment formed these beds of rock**

- The layers of rock are called **beds**.
- The boundaries between different layers are called **bedding planes**.

- The rock layer at the bottom of a sequence of layers is usually the oldest. Its sediment was probably laid down before the others.

- However, sometimes layers are put under great stress by powerful movements in the Earth's crust. They can be snapped, folded and even turned upside down sometimes.

## Layers of minerals

We can also get layers of minerals that were once dissolved in water forming layers of rock. For example, when ancient seas evaporated.

The salts come out of the sea water as solids in sequence. The least soluble solid **precipitates** out first as the water evaporates off. So the first solid (precipitate) is usually calcium carbonate.

About 90% of the water has to evaporate from sea water before common salt (sodium chloride) comes out of solution . This is thought to have happened when seas became cut off and surrounded by land.

- We call rocks formed like this **evaporites**.

### ◄ CHECKPOINT ►

Name two minerals that can be found as evaporates:

_____

# 8G6 Evidence from layers

We can work out what happened millions of years ago by looking at rock layers:

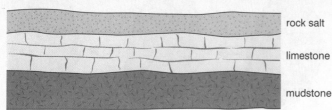

rock salt

limestone

mudstone

From these three layers we can say that a sediment of mud was probably laid down first, followed by the limestone, then the salt. That's because the oldest rocks usually come at the bottom of a sequence of layers.

## Layers from living things

Sometimes the layers of sediment found in rock can be formed from the remains of plants or animals. Chalk was made from the 'hard bits' of tiny sea plants called coccoliths.

- The sediments from living things **accumulate** (build up) in layers.

Coal is another rock made from sediments. It was formed millions of years ago from layers of plant material. The layers built up and were compressed, eventually turning into coal.

- You often find **fossils** in these rocks.

### ◄ CHECKPOINT ►

What is a coccolith?

_____

Which rock is formed from them?

_____

## Looking at rocks (8G1)

**1** Fill in the gaps in the following sentences:

Minerals are _____ elements or _____ found naturally. Most rocks are _____ of minerals.

There are two main types of rock _____: crystalline and _____

When the grains in a rock do not _____ the rock is _____, meaning it can _____ up water.

**2 a** Draw a diagram in the box to show a rock with a fragmental texture. Then explain why the rock is porous.

This rock is porous because _____

_____

_____

**b** Think of a model you could use to help explain the porosity of rocks to a pupil in Year 6. It should demonstrate interlocking and non-interlocking minerals.

_____

_____

_____

**3** Why do minerals have a chemical formula but rocks do not?

_____

_____

## Chemical weathering (8G2)

**1 a** Unscramble the letters to find the chemical name of the mineral contained in limestones:

licumac ratcabone _ _ _ _ _ _ _ _   _ _ _ _ _ _ _ _ _

**b** What does the mineral from part **a** form when it reacts with an acid?

_____

**c** What do we mean by the 'weathering' of a rock?

_____

**d** Why are some types of weathering called 'chemical weathering'?

_____

**2 a** Name three minerals we can find in granite:

_____

**b** Explain briefly how granite gets chemically weathered.

_____

_____

_____

## Physical weathering (8G3)

**1**  Describe what is happening in each stage of freeze/thaw weathering.

STEP 1

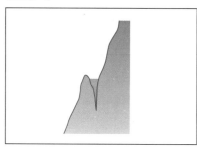

STEP 2

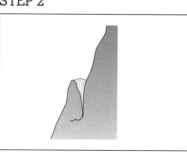

STEP 3

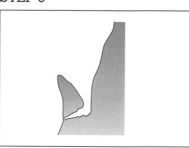

_____   _____   _____

_____   _____   _____

_____   _____   _____

_____   _____   _____

**2**  **a**  Give a similarity between physical weathering and chemical weathering.

_____

**b**  Give a difference between physical weathering and chemical weathering.

_____

## Fragments in motion (8G4)

**1**  Join the word to its definition with a line:

| Deposit |
| Erosion |
| Sediment |
| Weathering |
| Transport |

| The breakdown of rocks in nature |
| Rock fragments that are laid down in a particular place |
| Verb:  to lay down or settle an object in a place<br>Noun: the object that settles down |
| The movement of rock fragments from one place to another |
| The wearing away of rocks as they come into contact when moving over each other |

**2**  Fill in the gaps in the following sentences:

Weathered rock can be _____ by gravity, _____ , ice and

_____ .

The weathered pieces of rock become _____ , smoother and rounder as they are

_____ during movement from one place to another.

**1**　**a**　Explain how rock fragments can be sorted by size on a beach.

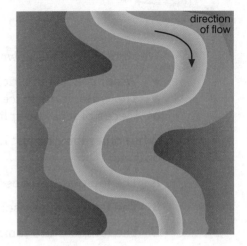

direction of flow

_____

_____

_____

_____

　　**b**　Here is a plan view of a meandering river (one that curves from side to side as it crosses the land).
Shade on the plan where you would expect bits of rock to be deposited.

**Evidence from fossils**

We can observe the sediments that were deposited millions of years ago in rocks exposed today.

Some sediments contain fossils. We can use the fossils, not only to build up a picture of species that are now extinct, but also to help age rocks found in different places. Fossils of the same species must be in rocks that were formed at about the same time in the history of the Earth.

Fossils can also give us clues about what conditions were like when the rocks were formed. For example, the tropical fern-like fossils found in the coal seams under Britain suggest that our climate was very different millions of years ago. In fact, most scientists believe that Britain was part of a land mass that has moved slowly from near the Equator to its present position over time.

**1**　Read the passage above and give two ways in which fossils help us to build up a picture of the Earth's history.

_____

_____

_____

_____

**2**　Draw a labelled diagram, in the box, of the layers you would expect to find from the evidence in this passage:

Millions of years ago a sediment of sand built up on the bed of an ancient sea. The landscape changed: the next layer to build up came from more turbulent conditions as pebbles and stones were deposited. Then, following more changes, a period of calm arrived. Sea levels were high now and the only sediment coming to rest on this part of the seabed was the remains of tiny plants called coccoliths.

# APPLY YOUR KNOWLEDGE

**(SAT-style questions)**

**1** Weathering breaks down rocks. This can happen when:

- Water collects in cracks in rocks.
- The water turns to ice at its freezing point and expands.
- Pieces of rock get split off.

What other things must happen during this type of weathering? Tick two of these options:

**A** The temperature does not change.

**B** The temperature rises above 0 °C.

**C** The temperature remains below 0 °C.

**D** Expansion forces all the water out of the cracks.

**E** Expansion causes the cracks to widen.

**F** Expansion forces the cracks to close up. (2)

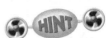

HINT: This type of physical weathering is called freeze/thaw!

**2** There are potholes, caves and caverns in limestone regions:

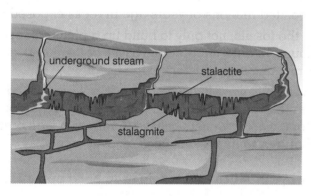

underground stream
stalactite
stalagmite

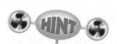

HINT: What breaks down the limestone causing chemical weathering? How can this reaction form a cave?

**a)** Explain how chemical weathering can form caverns in the limestone.

______________________________

______________________________

______________________________ (2)

**b)** Explain how stalactites form in the caves.

______________________________

______________________________ (2)

HINT: Think what will happen to the dissolved substances from the limestone when water evaporates off from drops in the cave.

**c)** Limestone can absorb water. It is a porous rock.
Describe how this causes the physical weathering of limestone.

______________________________

______________________________ (2)

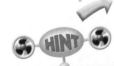

HINT: Think of what happens when water freezes.

Pip and Benson carry out an investigation into the porosity of rock:

- They have a piece of sandstone and a piece of basalt rock.
- They weigh each piece of each rock when dry and then again after soaking both in water.

Here are their results:

| Rock | Mass when dry (g) | Mass after soaking (g) | Increase in mass (g) | Percentage increase in mass |
|---|---|---|---|---|
| sandstone | 200 | 230 | | |
| basalt | 50 | 51 | | |

a) Complete the table. (2)

b) Why was it necessary to work out the percentage increase in mass?

_____

_____ (1)

c) i) Which rock is more porous? _____ (1)

   ii) How do the results show this?

   _____

   _____ (1)

Remember that porous rocks can soak up water.

d) The texture of basalt is called crystalline. It is made up of interlocking crystals.
   Sandstone has a fragmental texture. Its grains do not interlock.
   Explain the difference in the results between the two rocks.

How does water get inside the rock?

   _____

   _____

   _____ (2)

e) Pip and Benson had a third unknown type of rock, labelled C, to test.
   Rock C had a mass of 150 g when dry, and 180 g after soaking in water.

   i) What type of texture does rock C have? _____ (1)

   ii Put the three rocks – sandstone, basalt and C –
      in order of porosity.

   _____ most porous

   _____

   _____ least porous (1)

   Work out the percentage increase in mass of Rock C and compare that to the results in the table for sandstone and basalt.

f) Give one way in which Pip and Benson could improve the reliability of their results.

   _____

   _____ (1)

How could they make it more likely that someone repeating their experiment would get the same results?

# The rock cycle

## 8H1 Sedimentary rocks

### How do sediments turn to rock?

Over time, layers of sediment are deposited and build up. The individual bits of rock become a layer or bed of rock:

- The pressure builds up as layer upon layer of sediment is deposited.
- This squeezes water out from the gaps between the grains of sediment.
- Under this pressure, the edges of the grains can fuse together.
- This is called **compaction**.

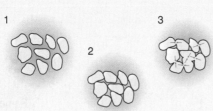

Another process also helps the sediments to form rock:

- Water that passes between the gaps in grains can evaporate, leaving behind any solids that were in solution.
- The solid that comes out of solution acts like a glue or cement, sticking the grains of sediment together.

- This is called **cementation**.
- Rocks formed like this are called **sedimentary rocks**.

### Sedimentary rocks

We can say that sedimentary rocks:

- are formed by the processes of compaction and cementation
- are usually porous (absorb water)
- usually have fragmental textures (grains that don't interlock)
- can contain fossils.

> **◄ CHECKPOINT ►**
>
> Fill in the gaps in the following sentences:
> Sedimentary rock is formed by the processes of
>
> _____ and _____.
>
> The _____ increases as _____ of sediment and rock build up above.
>
> This cause the edges of _____ to fuse _____, and water is squeezed out.

## 8H2 Looking at limestones

Different types of limestone were formed under different conditions.

### Shelly limestones

- Shelly limestones formed from sediments of shells accumulated on ancient seabeds.
- Sea creatures extract calcium carbonate to form their shells. These are often broken up by water currents before settling as a sediment.
- In clear waters, these shells will be the main sediment collecting on the seabed. Then we find a high proportion of calcium carbonate in the limestone.
- If there are other sediments mixed with the shelly bits, then the proportion of calcium carbonate is reduced.

**Brown limestone containing minerals of iron as well as calcium carbonate**

### Chalks

We have already seen how chalk was formed slowly over millions of years from the remains of coccoliths. These were tiny sea plants.

There is about 98% calcium carbonate in chalk. This means that there were few other sediments settling on the seabed at the same time as the coccoliths.

One theory suggests that sea levels were very high at the time when chalk was laid down. So there was not much land exposed.

This meant there was very little sediment carried to the sea by rivers. At present, similar deposits of sediment are only found at the bottom of oceans well away from land.

> **◄ CHECKPOINT ►**
>
> What different types of living things were shelly limestone and chalk made from?
>
> _____

# 8H3 Metamorphic rocks

Sometimes rocks are subjected to very high temperatures and/or pressures.

When this happens, chemical reactions take place in the solid rock. New minerals will form and recrystallisation occurs.

No new elements can be added within the rock. However, those already in the original minerals are re-arranged to make the new minerals.

- The new rock formed is called a **metamorphic rock**.
- Slate is formed under high pressure, caused by movements in the Earth's crust or very deep burial under many layers of rock.
- The new minerals in slate are all lined up in one direction. The crystals grow at right angles to the pressure.
- Rocks are also subjected to extreme heat from molten rock, called **magma**.
- The magma rises towards the surface in areas where we find volcanoes. The earth movements that build mountains also generate great heat. The rocks that are changed get very hot but they do not melt.
- Marble can be formed by the action of heat on limestone or chalk.

mudstone

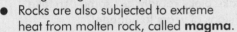

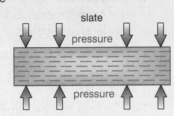

slate
pressure
pressure

In general, metamorphic rocks are:

- Made of crystals that are often too small to see with the naked eye.

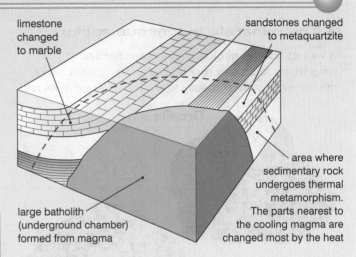

limestone changed to marble

sandstones changed to metaquartzite

area where sedimentary rock undergoes thermal metamorphism. The parts nearest to the cooling magma are changed most by the heat

large batholith (underground chamber) formed from magma

- Made of crystals that are usually interlocking, so the rocks are non-porous.
- Often found with bands of minerals running through the rock.
- Without fossils, except distorted ones in 'low grade' metamorphic rock, such as slate.

◄ CHECKPOINT ►

Fill in the gaps in the following sentences:
New rocks that have been formed by the action of _____ and/or heat (without _____ the rock) are called _____ rocks. You can often see _____ of minerals running through the rock.

# 8H4 Igneous rocks

We saw in 8H3 how molten rock, called **magma**, can rise towards the Earth's surface.

Sometimes it actually escapes from the surface, as in a volcano.

The molten mixture of materials that breaks through the surface is called **lava**.

**This lava contains a mixture of minerals and gas**

- When molten rock (magma or lava) solidifies, **igneous rock** forms.
- As magma or lava cools down, the interlocking crystals, which make up igneous rock, form.
- The **slower** the rate of cooling, the **larger** the crystals that form.

**Granite's large crystals formed as molten rock cooled slowly under the ground, surrounded by thick layers of rock. It is called an *intrusive* igneous rock.**

**Basalt's small crystals were formed as molten rock cooled quickly at or near the surface. The crystals are so small that you need a microscope to see them. Much of the ocean floor is made from basalt. It is called an *extrusive* igneous rock.**

◄ CHECKPOINT ►

Why is granite called an 'intrusive' igneous rock and basalt 'extrusive'?

_____

_____

# 8H5 Recycling rock

## The minerals in igneous rocks

As well as texture, geologists also use chemical composition to classify rocks. In intrusive igneous rocks, with larger crystals, it is easy to see the different minerals.

### Granite

### Gabbro

- Granite is a light-coloured igneous rock. It is rich in the elements silicon and oxygen (present in silicate minerals). Granite is said to be a silica-rich rock, containing the mineral quartz.

- On the other hand, gabbro is much darker in colour. It is rich in minerals that contain iron, called **mafic** minerals.

- The minerals containing iron are denser, so gabbro has a higher density than granite.

- Basalt is another igneous rock high in iron-bearing minerals. It is like gabbro in its mineral content. However, it was cooled down much more quickly when it formed. Its crystals are too small to see with the naked eye.

In general igneous rocks are:

- made of interlocking crystals
- hard and non-porous.

However, if they are formed from a spray of lava, they can land as ash.

## The rock cycle

Here are the three main types of rock:

- Sedimentary
- Metamorphic
- Igneous.

They are involved in a long cycle of change.

Some changes take place rapidly, but some can take thousands or millions of years to happen:

- The formation of igneous rocks on the Earth's surface is a quick process. Lava hitting cold sea water solidifies in seconds.

- However, the burial, compaction and cementation of sedimentary rocks is a very slow process. It can take millions of years.

The changes can be summarised in the **rock cycle** (see below).

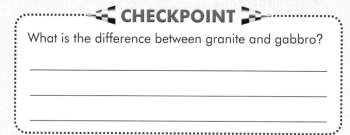

### ◄ CHECKPOINT ►

What is the difference between granite and gabbro?

_____

_____

_____

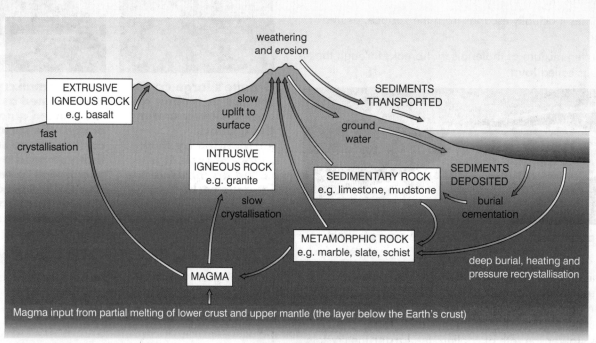

**The rock cycle**

## Sedimentary rocks (8H1)

**1**  **a**  Sometimes fine particles of sediment (such as clay or silt) settle between pebbles. Under pressure from the layers above, the fine sediment can join together. This helps to bind the pebbles into a rock called **conglomerate**.

Is this description an example of compaction or cementation? _____

**b**  What else is likely to happen as the conglomerate rock is formed, as well as the process described in part **a**?

_____

**2**  If a rock has a 'fragmental texture', what does this mean?

_____

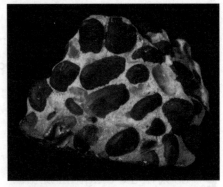

**Conglomerate is a sedimentary rock made from a sediment containing pebbles**

## Looking at limestones (8H2)

**1**  Read this information, then answer the questions that follow:

Coral reefs are made up of the shells of sea creatures that fasten themselves to the bottom of the sea in warm conditions. These reefs build up over time and form a very hard rock called **reef limestone**.

Sometimes calcium carbonate comes out of solution when some sea water evaporates off. Sometimes this happens in calm waters, for example in a lagoon. Then the tiny crystals of calcium carbonate that form will sink to the seabed and form a fine white mud (called **micrite**). Under compaction this can form rock. The sediment is often mixed with mud so the rock formed is dark grey with very fine grains, called **lime 'mudstone'**.

The fine white mud of calcium carbonate can also coat other bits of sediment that roll gently across it. In effect you get tiny snowballs with an outer layer of calcium carbonate. These can go on to form another type of limestone (**oolitic limestone**).

**Corals contain calcium carbonate in their shells**

**a**  Name three types of limestone mentioned in the passage above.

_____

**b**  What is 'micrite'?

_____

**Oolitic limestone**

**c**  Calcium carbonate is white. So why are some types of limestone grey in colour?

_____

**d**  You find a large pebble with a thin, white outer layer of calcium carbonate. How might the outer layer have been formed?

_____

_____

## Metamorphic rocks (8H3)

**1**   Unscramble these letters to find the names of the rocks:

**a**   **n o t m i l e s e** can turn into **r e b l a m**

**b**   **s u m n o t e d** can turn into **t a s e l**

**c**   Using your answers to part **a**, fill in this table:

| Example of sedimentary rock | Example of metamorphic rock |
|---|---|
|  |  |
|  |  |

**2**   Look at the diagram below:

**a**   Shade in the area on the diagram where you would expect to find metamorphic rock.

**b**   Explain what happened to form the metamorphic rock marked on the diagram in part **a**.

_____

_____

_____

_____

_____

_____

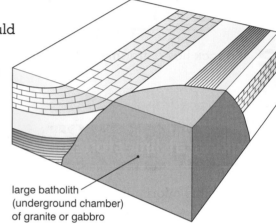

large batholith
(underground chamber)
of granite or gabbro

## Igneous rocks (8H4)

**1**   Fill in the gaps in the following sentences:

When molten rock (_____ or lava) cools down, it usually s _____ into

crystals of _____ rock.

If the _____ cools down _____, deep underground, then we get

_____ crystals.

Gr _____ is an example of such an _____ igneous rock.

If it cools down _____ at or near the surface, _____ crystals form.

B _____ is an example of one of these _____ igneous rocks.

**2**   The igneous rock formed at the edge of a batholith has smaller crystals than the rock in the centre. Explain this.

_____

_____

**1**   Identify the rocks below as igneous, metamorphic or sedimentary:

a   ROCK A: It is made from plate-like crystals all lined up in the same direction. The rock fragment has parallel flat sides where it has been cleaved.

ROCK A is _____

An example of this type of rock is _____

b   ROCK B: There are three different types of interlocking crystal arranged randomly in this hard rock.

ROCK B is _____

An example of this type of rock is _____

c   ROCK C: There are particles of sand visible, held together by an orangey brown mineral. Bits of sand crumble off the surface of the rock quite easily.

ROCK C is _____

An example of this type of rock is _____

**2**   Label the rock cycle below:

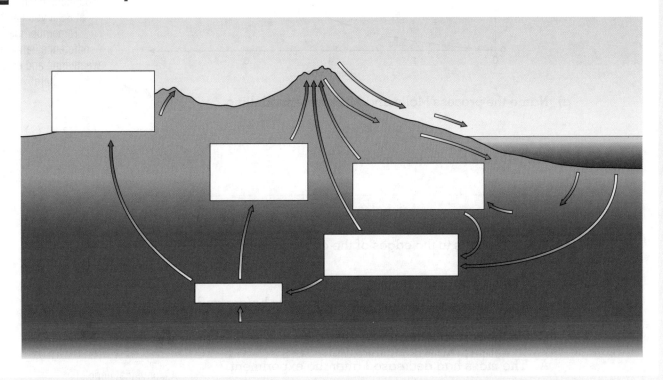

**1** Molly and Mike were investigating how rocks can be worn down:

- They made six cubes from plaster of Paris.
- They weighed the cubes then put them in a tin can with a lid.
- They shook them for 30 seconds then weighed the six largest blocks again, making sure no bits were lost from the can.
- They replaced the blocks in the can and repeated this several times.

Here is a graph of their results:

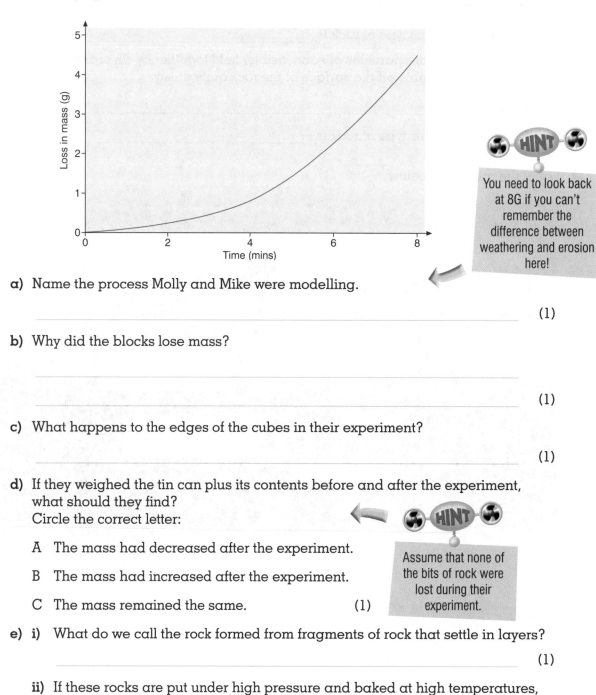

**HINT**

You need to look back at 8G if you can't remember the difference between weathering and erosion here!

**a)** Name the process Molly and Mike were modelling.

_____  (1)

**b)** Why did the blocks lose mass?

_____

_____  (1)

**c)** What happens to the edges of the cubes in their experiment?

_____  (1)

**d)** If they weighed the tin can plus its contents before and after the experiment, what should they find?
Circle the correct letter:

**HINT**

Assume that none of the bits of rock were lost during their experiment.

A  The mass had decreased after the experiment.

B  The mass had increased after the experiment.

C  The mass remained the same.  (1)

**e) i)** What do we call the rock formed from fragments of rock that settle in layers?

_____  (1)

**ii)** If these rocks are put under high pressure and baked at high temperatures, what type of rock forms?

_____  (1)

**2** Look at these five steps:

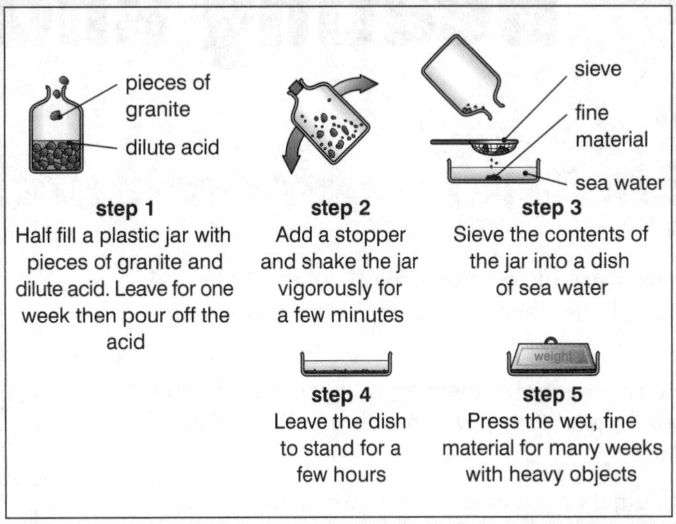

a) What type of rock is granite? (Circle the correct letter.)

   A  igneous

   B  metamorphic

   C  sedimentary    (1)

b) What process is represented in Step 1?

_____ (1)

c) Which process is modelled in Step 2?

_____ (1)

**step 1**
Half fill a plastic jar with pieces of granite and dilute acid. Leave for one week then pour off the acid

**step 2**
Add a stopper and shake the jar vigorously for a few minutes

**step 3**
Sieve the contents of the jar into a dish of sea water

*sieve*
*fine material*
*sea water*

*pieces of granite*
*dilute acid*

**step 4**
Leave the dish to stand for a few hours

**step 5**
Press the wet, fine material for many weeks with heavy objects

*weight*

d) Rivers carry bits of granite from the mountains to the sea. Give **two** ways that the bits of granite change on this journey.

_____ (2)

**HINT**

What will happen to the size and the shape of the bits of granite?

e) Step 4 and Step 5 represent **two** parts of the rock cycle. What are they?

_____ (2)

f) Name a rock that could have formed as modelled in the diagram.

_____ (1)

**3** The list below describes some processes that occur in the rock cycle:

A  Layers of new minerals form as the mudstone is squeezed.
B  Deep in the Earth's crust, rocks can be subjected to high temperatures and pressures. New crystals can then be formed in bands running through the rock.
C  Grains collect on the seabed.
D  As molten magma cools deep underground, large crystals form.
E  Grains of sediment get cemented together as they are buried under more and more sediment.
F  When molten lava erupts into the sea, a glassy rock solidifies.

a) Give the letters of two processes that describe the formation of metamorphic

   rock. _____ (2)

b) Give the letters of two processes that describe the formation of igneous

   rock. _____ (2)

c) Which two letters could lead to the formation of sandstone? _____ (2)

d) The glassy rock described in F is called 'obsidian'. Why does it contain no crystals?

_____ (1)

e) Which letter describes the process of deposition?

_____ (1)

**HINT**

**Obsidian** is formed by very rapid cooling, therefore

## 811 Taking temperatures

The **temperature** of something tells you how hot it is. If one object is hotter than another is, its temperature is higher.

We can sense how hot something is by touching it. The skin of our fingers has **nerve cells** that can sense the temperature.

However, to get an accurate measurement of temperature, we need to use a **thermometer**. There are many different types of thermometer:

- Alcohol-in-glass
- Liquid crystal
- Electronic.

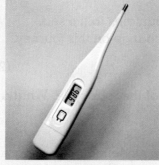

In science, we measure temperatures on the **Celsius scale**. For example, we say:

Pure water freezes at 0 °C (zero degrees Celsius).

It's useful to remember some important temperatures:

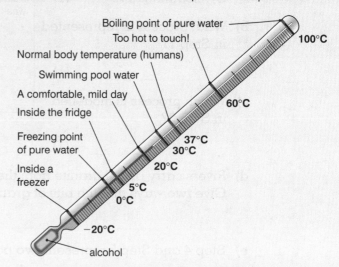

Boiling point of pure water — 100°C
Too hot to touch!
Normal body temperature (humans)
Swimming pool water
A comfortable, mild day — 60°C
Inside the fridge
Freezing point of pure water — 37°C / 30°C
Inside a freezer — 20°C
5°C
0°C
−20°C
alcohol

**◄ CHECKPOINT ►**

Which is normally at a higher temperature, the water in a swimming pool, or a person swimming in it?

_____

## 812 Getting hotter, getting colder

You have probably noticed that hot things tend to cool down, and cold things tend to warm up. For example:

- A hot drink **cools down** until it is the same temperature as the room.
- A cold drink **warms up** until it is the same temperature as the room.

As a hot drink cools down, it is losing energy. Energy escapes from hot objects, and spreads into colder ones.

In a similar way, cold objects get hotter, because energy spreads into them from their warmer surroundings.

When energy moves from a hotter place to a colder place, we say that **heat energy** is flowing.

- **Heat energy always flows from a hotter place to a colder one.**

**Take care!** We don't say that cold energy flows from cold objects into hotter ones!

If you want to predict which way heat energy will flow, look at the temperatures. It always flows from a hotter object to a colder one.

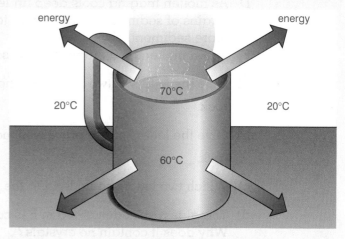

energy — energy
20°C — 70°C — 20°C
60°C

**◄ CHECKPOINT ►**

Fill in the gaps in the following sentences:
You put a lump of ice (0 °C) into a glass of lemonade

(20 °C). Heat energy will flow from the _____

into the _____.

# 813 Conduction, insulation

## Thermal insulators

Heat energy flows from hotter to colder places. We can make it flow more slowly by using insulating materials.

Here are some examples of materials that are good **thermal insulators**:

- Wool, cotton, nylon
- Plastic and glass
- Air.

Heat energy doesn't flow easily through these materials. That's why we wear woolly clothes on a cold day, so that we don't get cold.

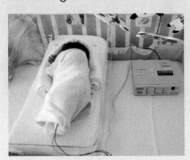

**This tiny baby needs to be well insulated to keep it warm**

## Thermal conductors

Metals are good **thermal conductors**. Heat energy can flow easily through metals. Copper is an especially good thermal conductor.

If you touch an object that is at room temperature, you may be able to judge whether it is an insulator or a conductor:

- If it feels warm, it's an insulator.
- If it feels cold, it's a conductor.

When you touch a piece of metal, heat energy flows out of your finger into the metal. Your finger cools down, and the nerves in your finger tell you that the metal is cold.

When you touch a piece of plastic, energy cannot flow out of your finger, because plastic is an insulator. So your finger stays warm, and the plastic feels warm.

**◀ CHECKPOINT ▶**

Cross out the incorrect words:
Woollen clothes feel warm/cool because wool is a good thermal conductor/insulator.

# 814 Expanding and explaining

You have probably used an alcohol-in-glass thermometer to measure the temperature of some water. If you heat the water, the alcohol moves up inside the glass tube. This happens because the alcohol expands as it gets hotter.

Most materials **expand** when they get hotter, and **contract** when they get cooler.

A thermometer makes good use of expansion, but expansion can also be a problem. For example, railway lines may expand and buckle on a hot day.

When a material is heated, its particles vibrate more and more. They push each other apart, so that the material gets bigger.

If we think about vibrating particles, we can also explain how heat energy is conducted. The particles that are vibrating most share their energy with particles that are vibrating less.

cold          hot

**◀ CHECKPOINT ▶**

What word is the opposite of 'contraction'?

_____

# 815 Radiation and convection

**Radiation** is another way that heat energy escapes from a hot object. The Sun is very hot, so it radiates energy very fast. Even people radiate energy, but we are much cooler than the Sun, so we radiate much more slowly.

Another name for heat energy spreading out in this way is **infra-red radiation**.

Heat energy can also travel by **convection**. For example, when sunlight warms the air, it expands and rises. (It rises because it is less dense than the surrounding air.) Cold air flows in to replace it. This is called a **convection current**. The wind and ocean currents are caused by convection currents.

**◀ CHECKPOINT ▶**

Fill in the gaps in the following sentence:
Heat energy travels in three different ways:
_____tion, _____tion and _____tion.

# 816 Conserving energy

It's important not to waste energy. Energy escapes from our houses in different ways:

- By **conduction** through the walls, floors and windows.
- By **convection**; warm air rises above the house; draughts are convection currents.
- By **radiation**, because the house is warmer than its surroundings.

If we waste energy, we are wasting money. We are also damaging the environment and helping to cause climate change. So it is important to **conserve** energy.

Houses can be designed to conserve energy:

- Add **insulation** to the walls, floor and roof.
- Double-glaze the windows.
- Fit draught excluders and thick curtains.

**Animals that live in cold places need to conserve energy too**

### ◀ CHECKPOINT ▶

Complete this sentence:
The rabbit in the photograph has thick _____ to conserve energy.

# 817 Changing state

Freezing, boiling, melting, condensing are **changes of state**.

If you heat a solid, it may melt. As it melts, its temperature stays steady even though you are heating it. Similarly, the temperature of a liquid remains steady when it is boiling.

- The **freezing point** is the temperature at which a liquid freezes and becomes a solid.

- The **boiling point** is the temperature at which a liquid boils and becomes a gas.

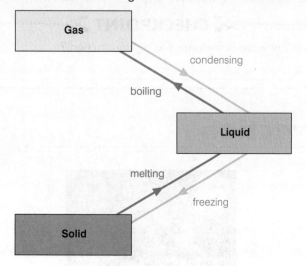

We can understand changes of state by thinking about the particles of which a material is made.

**Gas:** Particles are separated, free to move about

**Boiling:** Particles break completely free of their neighbours

**Liquid:** Particles close together, can move around a bit

**Melting:** Particles start to break free of their neighbours

**Solid:** Particles close together, can't move about

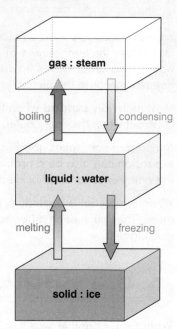

### ◀ CHECKPOINT ▶

Complete:
Melting point is the same as _____ _____.

## Taking temperatures (8I1)

**1** What quantity tells us how hot something is? _____

**2** What scale do we use to measure this quantity? _____

**3** Some thermometers are made from a glass tube containing mercury. Name another liquid that is used in thermometers. _____

**4** What is normal body temperature for a human being? _____

**5** At what temperature does pure water boil? _____

**6** The table on the right lists five temperatures, A-E. Put them in order, from lowest temperature to highest. Fill in the boxes below.

| A | normal body temperature |
|---|---|
| B | freezing water |
| C | a cool day |
| D | inside the Sun |
| E | boiling water |

Lowest ☐ ☐ ☐ ☐ ☐ **Highest**

## Getting hotter, getting colder (8I2)

**1** Cross out the incorrect words:

Heat energy flows from a hotter/colder object into a hotter/colder object.

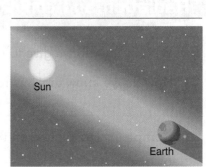

**2** Look at the picture of the hot drink:

**a** What is the temperature of the drink? _____

**b** What is the temperature of its surroundings? _____

**c** Why does energy flow out of the drink, into its surroundings? _____

**3** During the day, the Earth warms up. At night, it cools down. We can understand why by thinking about temperatures:

- The Sun is very hot.
- The Earth is quite warm.
- Space is very cold.

**a** On the drawing, add arrows to show how heat energy flows.

**b** Write a sentence to explain why it gets cold at night.

_____

_____

_____

## Conduction, insulation (813)

1  What is the opposite of a 'thermal conductor'? _____

2  In the box on the right:

- Underline all the good thermal conductors.
- Put a star next to the best thermal conductor.
- Put a ring around all the poor thermal conductors.

| Copper | Air |
|---|---|
| Polythene | |
| Glass  Steel | Wool |
| Wood  Aluminium | |

3  Explain why a saucepan is usually made of metal, but its handle is made of plastic.

_____

_____

## Expanding and explaining (814)

1  Complete: An alcohol-in-glass thermometer makes use of the fact that liquids _____ when they get hotter.

2  Choose words or phrases from the list on the right to complete these sentences:

When a solid is heated, its particles vibrate _____.

This means that the particles _____

and so the solid _____.

get bigger
expands
take up more space
more
less
explodes

3  The diagram on the right shows a metal rod. It is hot at one end, and cold at the other.

a  Add an arrow to the diagram to show how heat energy flows along the rod.

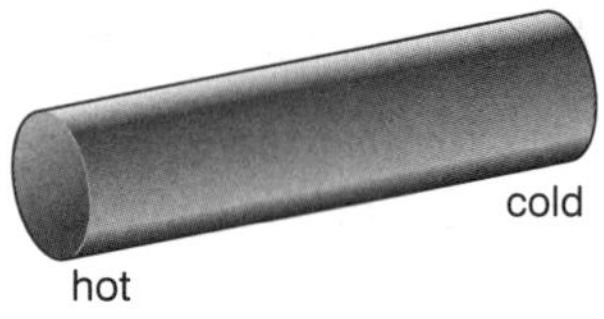

b  Complete: The vibrating particles jostle their neighbours, and _____ their energy with them.

This is how heat energy travels by _____.

hot    cold

## Radiation and convection (815)

1  What name do we give to heat energy travelling by radiation? _____ radiation.

2  How do we know that this radiation can travel through empty space? (HINT: Think about the Sun.)

_____

3  Complete: The picture on the right shows how heat energy can be carried by convection _____.

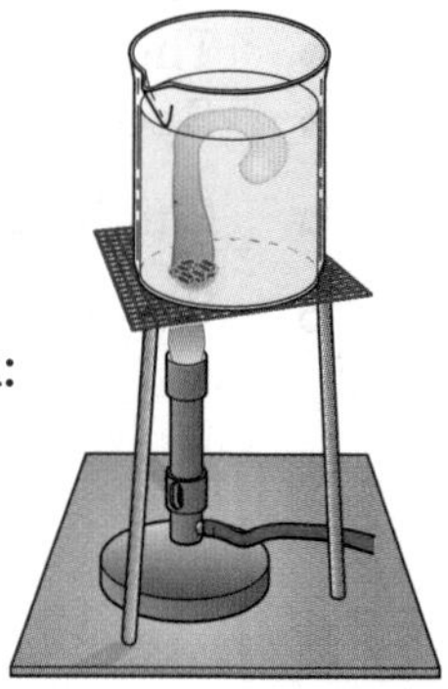

4  Fill in the gaps in the following sentences that show how we explain convection:

a  When a liquid is heated, it e_____.

b  Its d_____ becomes less, so it f_____ upwards.

c  Now c_____ liquid flows in to replace it.

## Conserving energy (816)

**1** In the list on the right:

   **a** Underline the word that means the **same as** conserving energy.

   **b** Cross out the word that means the **opposite of** conserving.

> wasting
>
> saving
>
> conducting
>
> transferring

**2** Complete: You can conserve energy by adding i_____ to your house. You will also save m_____.

**3** Heat energy can escape from a house in different ways. For each of these, write whether the energy is escaping by **conduction**, **convection** or **radiation**:

Through the solid floor: _____

By draughts: _____

From the warm roof and walls: _____ and _____.

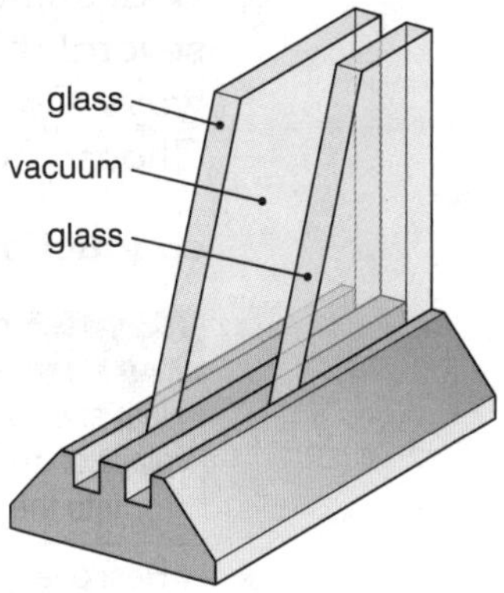

**4** Complete:
The picture shows that a double-glazed window has a vacuum between two panes of glass.

This means that heat energy can escape by _____ tion,

but not by _____ tion or _____ tion.

glass

vacuum

glass

## Changing state (817)

**1** Complete:
The three states of matter are: _____, _____ and _____

**2** Name these changes of state:

   **a** From solid to liquid: _____    **c** From liquid to solid: _____

   **b** From gas to liquid: _____    **d** From liquid to gas: _____

**3** Which states of matter are described here?

   **a** Particles close together, can move around: _____

   **b** Particles in fixed positions: _____

   **c** Particles far apart, moving freely: _____

**4** Which changes of state are described here?

   **a** Particles break completely free from their neighbours: _____

   **b** Particles stick tightly together, no longer free to move

     about: _____

**5** Pip heated some wax, so that it melted. She measured its temperature as it cooled, and the graph shows her results.

   **a** On the graph, mark the following regions: solid only, liquid only, solid and liquid.

   **b** What was the freezing point of the wax? _____

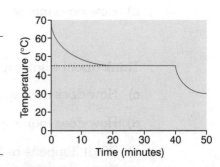

# APPLY YOUR KNOWLEDGE

**1** Molly was investigating how a beaker of hot water cooled down. The drawing shows her apparatus.

**a)** Name the instrument she used to measure the temperature of the water:

_____

(1)

**b)** Heat energy escaped from the beaker of water in several different ways; the table shows some of the ways. Complete the table by filling in the last column. Choose from:

**conduction     convection     insulation     radiation**

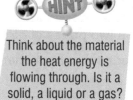

| How the heat energy escaped | Name for this |
|---|---|
| Heat energy passed through the glass of the beaker and spread into the top of the table. | |
| Heat energy was carried away by warm air rising above the beaker. | |

Think about the material the heat energy is flowing through. Is it a solid, a liquid or a gas?

(2)

**c)** After a while, Molly measured three temperatures:

- temperature of water in beaker = 20 °C
- temperature of air near beaker = 20 °C
- temperature of table under beaker = 20 °C

Predict whether the water would continue to get colder. Give a reason to support your answer.

_____

(2)

**2** In Pete's room, there is an electric heater. Heat energy is carried around the room by convection currents.

Air close to the heater is heated.

**a)** How does the temperature of the air change?

_____

(1)

**b)** How does the density of the air change?

_____

(1)

Think about what happens to the particles of the air when air is heated:

**c)** How does their speed change? _____ (1)

**d)** How does their separation change? _____ (1)

**e)** What happens as the warm air rises? _____ (1)

You need to know about the movement of the particles to explain why convection happens.

Benson was investigating the freezing of sea water. He put a plastic beaker of sea water into a freezer, together with a temperature sensor connected to a computer to record the temperature of the water.

The graph shows how the temperature of the water changed.

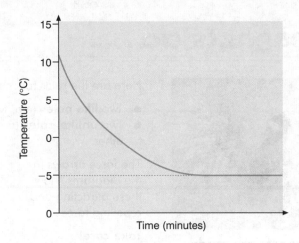

Use the graph to help you answer these questions:

You can find these answers from the graph.

a)  What was the temperature of the water when Benson put it

in the freezer? _____ (1)

b)  At what temperature did the water freeze? _____ (1)

The sea water froze at a different temperature to pure water.

c)  At what temperature does pure water freeze? _____ (1)

d)  Benson repeated the experiment with pure water instead of sea water. The temperature of the water was 20 °C when he put it in the freezer. On the axes below, draw a graph to show how you think his results would appear. (2)

The shape of the graph will be similar to the one at the top of the page. What temperature will it start at? What temperature will it end at?

# Magnets and electromagnets

## 8J1 What magnets do

Magnets have many different uses. They attract each other, and they attract **magnetic materials**.

Examples of magnetic materials are:

- Iron
- Cobalt
- Nickel
- Steel (most kinds).

**All magnets are made from magnetic materials**

Examples of non-magnetic materials are:

- Copper
- Aluminium
- Wood
- Water
- Plastic
- Stainless steel.

Many permanent magnets are in the shape of a bar. There is a **pole** at each end. The poles are where the magnet's attraction is strongest. They are called the **north pole** and the **south pole**.

Here are the rules for magnets:

- Two **like poles** (e.g. two south poles) **repel** each other.
- Two **unlike poles** (a north and a south) **attract** each other.

The force arrows in the diagrams show these attracting and repelling forces.

**Take care!**

- If two pieces **attract** each other, you can't be sure that they are both magnets. One might be a magnet, and the other a piece of unmagnetised metal.
- If they **repel** each other, you can be sure that they are both magnets.

**CHECKPOINT**

Cross out the incorrect word:
All metals are magnetic materials: True/false

## 8J2 Making and testing magnets

You can use a permanent magnet to magnetise a piece of iron or steel:

- Stroke the steel using one pole of the magnet.
- Always stroke it using the same pole.
- Always stroke it in the same direction.

Keep stroking to make the steel a stronger magnet. After a while, it won't become any stronger.

There are several ways to demagnetise a magnet: for example, hitting it with a hammer, or heating it.

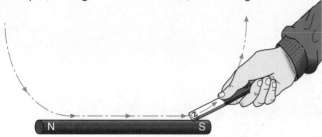

Here are two ways to test the strength of a magnet:

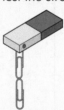

Hang paperclips from one end of the magnet. (Each clip attracts the one below it.) The longer the chain, the stronger the magnet.

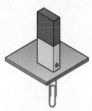

Put layers of cardboard between the magnet and the paperclip. A stronger magnet will be able to attract the clip through more layers of cardboard.

- The paperclips become magnetised, so that they can attract each other.
- The magnetic force can pass through cardboard. (It can't pass through a magnetic material.)

**CHECKPOINT**

Cross out the incorrect word:
Can the force of a magnet pass through air? Yes/no

**Compasses** are used for **navigation**. They can also be used for investigating magnets.

A compass has a magnetic needle that is free to turn around; one end points north, and the other points south:

- The end that points north is called the **north-seeking pole** (or north pole).

- The end that points south is called the **south-seeking pole** (or south pole).

A magnet will attract any magnetic material that is nearby. We say that the magnet is surrounded by a **magnetic field**, and we draw magnetic field lines to show the shape of the field.

The magnetic field lines come out of the north pole of the magnet and go into its south pole.

Here are two ways to show up a magnetic field:

- Use iron filings

- Use a small plotting compass.

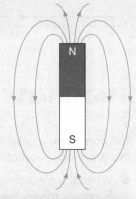

**Compasses have been used for centuries to help people find their way around, using the Earth's magnetic field**

The Earth has a magnetic field. That's why a compass needle points north–south. It's as if there were a giant bar magnet inside the Earth, with one pole near the North Pole and one near the South Pole. The Earth's magnetic field lines are similar to the field lines of a bar magnet.

**Take care!**

- There's a **south** magnetic pole near the **North** geographical pole.

- There's a **north** magnetic pole near the **South** geographical pole.

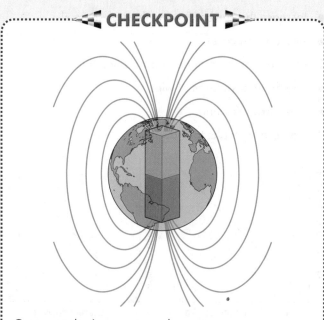

**◄ CHECKPOINT ►**

Cross out the incorrect word:
The north/south pole of a magnet points towards the geographical North.

# 8J4 Making an electromagnet

- A bar magnet is a **permanent magnet**. It stays magnetic all the time.

- Another type of magnet is an **electromagnet**. It needs electricity to make it magnetic.

- An electromagnet is made of a coil of wire, called a **solenoid**. When an electric current flows through the coil, it is magnetised. When the current is switched off, the magnetic field disappears.

- An iron **core** inside the coil makes the field much stronger.

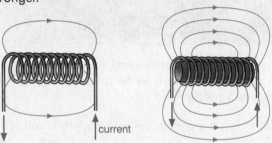

current

The magnetic field of an electromagnet is similar to the field of a bar magnet. One end is a north pole, and the other end is a south pole.

If the current is reversed (so that it flows through the coil in the opposite direction), the poles are reversed.

Electromagnets have many uses:

- In door bells and door bolts
- In automatic switches
- In cranes in scrap yards.

(There are also electromagnets in electric motors and generators.)

**◄ CHECKPOINT ►**

What must flow if an electromagnet is to work?

_____

# 8J5 Explaining electromagnets

There are several ways of making an electromagnet stronger:

- Use more turns of wire
- Increase the current flowing through the coil
- Add an iron core.

The strength of the magnet is proportional to:

- The number of turns of wire
- The current flowing.

We can draw graphs to show this:

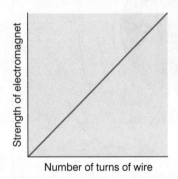

Number of turns of wire          Current flowing

Whenever an electric current flows, there is a magnetic field around it:

- By using a longer wire, we can make a stronger magnetic field.

- By making the wire into a coil, we can concentrate the magnetic field.

The core of an electromagnet is made of iron, because this is a magnetic material:

- When the electromagnet is switched on, the core becomes magnetised, so that the field is much stronger.

- When the current is switched off, the core loses its magnetism – it is demagnetised.

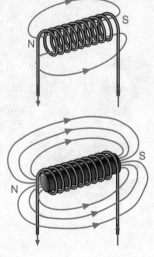

**◄ CHECKPOINT ►**

Cross out the incorrect numbers:
An electromagnet with twice as many turns and three times as much current will be 2 / 3 / 4 / 5 / 6 times as strong.

## What magnets do (8J1)

**1** In the box on the right:

- Underline all the magnetic materials.
- Cross out all the non-magnetic materials.
- Put a question mark by material which might be magnetic, but might not.

| Paper | Plastic | Steel |
| --- | --- | --- |
| | Iron | Copper |
| Water | Air | Nickel |
| | Wood | Cobalt |

**2** The force of a magnet is strongest at its p _____.

These are called n_____ and s_____.

**3** In an experiment, two magnets are placed end-to-end. They attract each other. Fill in the gaps:

a   When one magnet is turned the other way round, they _____ each other.

b   When the second magnet is turned round,

they _____ each other.

**4** The drawings show some magnets and some pieces of iron. For each pair, draw force arrows to show how they affect each other.

## Making and testing magnets (8J2)

**1** Look at the drawing on the right. It shows a simple experiment with a magnet, a length of cotton thread and a steel paperclip.

Which of the following things does this show? Cross out the incorrect words:

a   The magnetic force can pass through air: Yes/no/perhaps

b   Steel is a magnetic material: Yes/no/perhaps

c   Air is a magnetic material: Yes/no/perhaps

d   Cotton is a magnetic material: Yes/no/perhaps

**2** In the experiment, a piece of card was placed between the magnet and the clip.

a   What would you expect to see? _____

b   What would this show? _____

**3** The picture shows one way to test the strength of a magnet:

- You push the magnet towards the compass.
- A strong magnet will make the needle move from a greater distance.

Suggest how this method could be adapted to compare the strengths of two magnets:

_____

_____

_____

## Magnetic fields (8J3)

**1** Fill in the gaps in the following sentences:

It has a magnetised steel _____, which is free to turn around. It shows up the

magnetic field of the _____. What is it? A _____.

**2** Name two things that can be used to show up a magnetic field:

a   _____

b   _____

**3** The diagram on the right is incomplete. It shows the magnetic field around a bar magnet.

a   Label the magnet's north and south poles.

b   Add arrows to the magnetic field lines.

c   Below are some drawings of plotting compasses. Copy them onto the diagram, so that their needles are correctly showing the magnetic field.

## Making an electromagnet (8J4)

**1** An electromagnet is one type of magnet. What name is given to the other type (such as a bar

magnet)? _____ _____.

**2** To make a strong electromagnet, you need three things beginning with the letter c. Name them:

a   c_____

b   c_____

c   c_____

**3** The diagram shows an electromagnet. The top end is a north magnetic pole. On the diagram, draw field lines to show the magnetic field around the electromagnet. Remember to put arrows on, to show the direction of the field.

**4** What material is used for the core of an electromagnet?

_____

current

**1** The picture shows the things you might use to investigate the strength of an electromagnet. Add labels to the diagram to say what they all are.

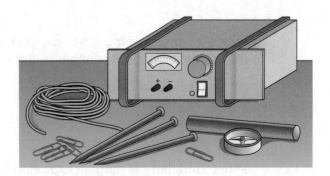

**2** Which of the following factors will make an electromagnet stronger? Tick the boxes.

☐ greater current

☐ use thicker wire

☐ add a plastic core

☐ add more turns of wire

☐ add an iron core

☐ draw more lines of force.

**3** The strength of an electromagnet is proportional to the number of turns of wire.

Which graph shows this correctly? Graph _____.

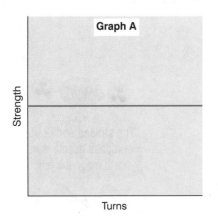

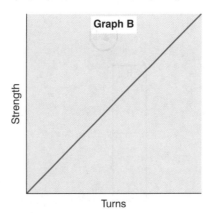

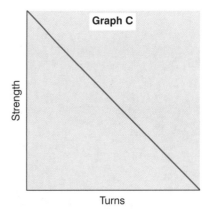

**4** Solve the clues to complete the puzzle:

**a** Flows through an electromagnet to make it work

**b** Another name for the coil of an electromagnet

**c** What an electromagnet's coil is made of

**d** The two ends of an electromagnet.

**e** Now make up your own clue for the word that appears in the shaded boxes:

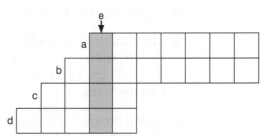

_____

# APPLY YOUR KNOWLEDGE

**(SAT-style questions)**

**1** Pete was investigating a bar magnet. His teacher gave him a second steel bar and asked him to find out whether or not it was a magnet.

**a)** Describe how he could do this. _____

_____

_____ (2)

Pete found that his magnet attracted the steel rod. He wrapped the magnet in a piece of cloth, and then brought it close to the steel bar:

**b)** Would the magnet still attract the bar? Explain your answer.

**HINT**

Which materials will a magnetic force pass through?

_____

_____ (2)

Then the teacher gave him some steel paperclips. Some were plastic-coated. Pete put the magnet into the paperclips.

**c)** What would you expect to happen? _____

_____ (1)

**2** Reese is investigating the magnetic field of a bar magnet. She places a small plotting compass near one of its poles, B, as shown in the diagram.

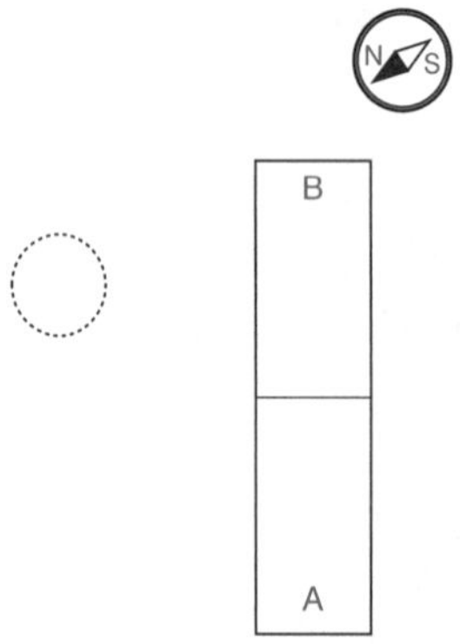

**HINT**

The **shaded** end of the compass needle is its north magnetic pole.

**a)** Is pole B a north magnetic pole or a south magnetic pole? _____ (1)

**b)** The diagram shows two other positions in which Reese placed the compass. On the diagram, mark the direction of the plotting compass needle in these two positions. (2)

**HINT**

Think about the pattern of the magnetic field lines of a bar magnet.

**c)** Reese hangs the magnet up using a length of string. The magnet is free to turn around. In which direction does Pole A point?

_____ (1)

**d)** Explain why the magnet points in this direction. _____

_____ (1)

**3** Molly made a coil of wire, which she connected up to a power supply and a switch to make an electromagnet. She placed two iron bars inside the coil, end-to-end. She held the ends of the iron bars. With the switch closed, she found it difficult to pull the two bars apart.

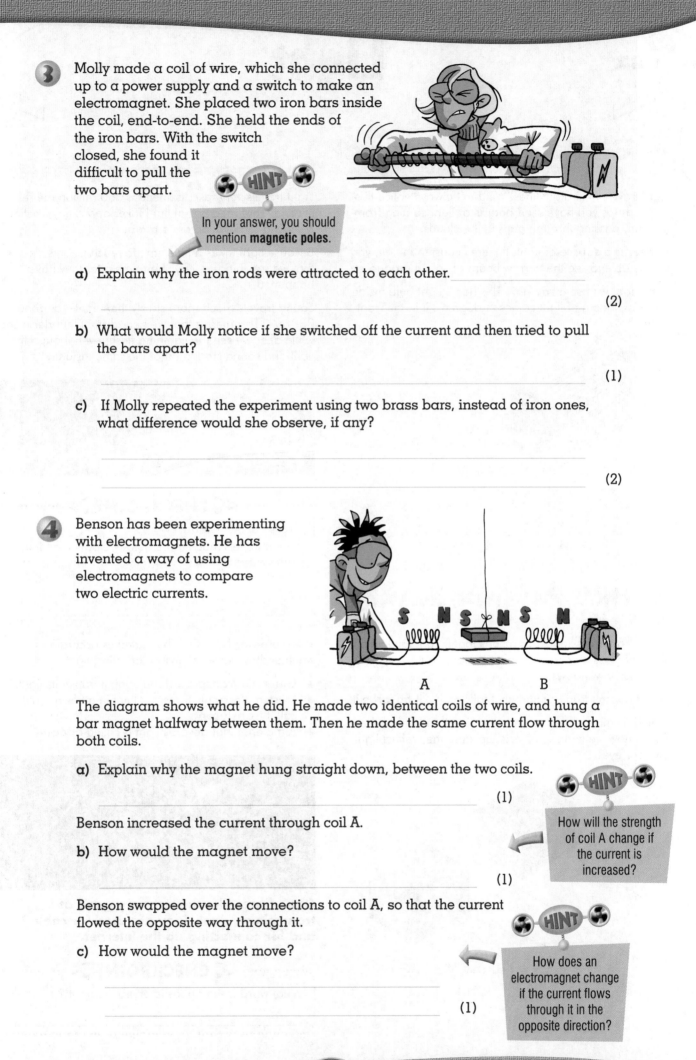

HINT

In your answer, you should mention **magnetic poles**.

**a)** Explain why the iron rods were attracted to each other.

(2)

**b)** What would Molly notice if she switched off the current and then tried to pull the bars apart?

(1)

**c)** If Molly repeated the experiment using two brass bars, instead of iron ones, what difference would she observe, if any?

(2)

**4** Benson has been experimenting with electromagnets. He has invented a way of using electromagnets to compare two electric currents.

S N S N S N

A                    B

The diagram shows what he did. He made two identical coils of wire, and hung a bar magnet halfway between them. Then he made the same current flow through both coils.

**a)** Explain why the magnet hung straight down, between the two coils.

(1)

Benson increased the current through coil A.

**b)** How would the magnet move?

HINT

How will the strength of coil A change if the current is increased?

(1)

Benson swapped over the connections to coil A, so that the current flowed the opposite way through it.

**c)** How would the magnet move?

HINT

How does an electromagnet change if the current flows through it in the opposite direction?

(1)

# Light

## 8K1 How light travels

Light travels in straight lines. We don't always notice this, but perhaps you have seen **beams** of light coming from the Sun, passing through gaps in the clouds.

A **laser** is a **source** of light. If there is dust in the air, you will be able to see the narrow beam of light it produces.

In the lab, we use a **ray box**. This has a light bulb inside; a narrow beam of light comes through a slit in the front of the box.

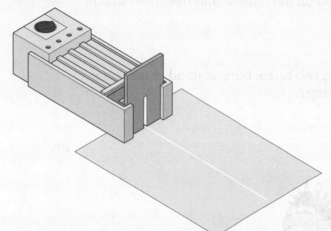

Light travels very fast. Its speed is 300 million metres per second. That means that light takes only a tiny fraction of a second to travel across a room.

Because light takes a little time to reach us, this means that we see things a very short time after they have happened.

Sound travels much more slowly than light – at about 330 m/s. This means that we hear distant thunder a short while after we see the lightning flash, even though the light and sound started off at the same instant.

light – very fast
sound – much slower

**◄ CHECKPOINT ►**

Cross out the incorrect word:
Sound travels much more quickly/slowly than light.

## 8K2 Passing through

Glass **transmits** light. This means that if light falls on a sheet of glass it passes straight through. That's why we use glass in windows.

Wood doesn't transmit light that falls on it – it **absorbs** it.

Materials may **reflect** some of the light that falls on them. That's how we can see objects, because they reflect light into our eyes.

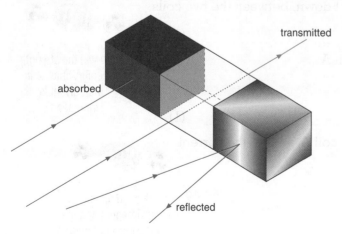

transmitted

absorbed

reflected

There are words to describe materials according to whether they transmit, absorb or reflect light:

- Glass is a **transparent** material; it transmits light.
- Frosted glass only transmits light in a blurry way; we call it **translucent**.
- A material that absorbs light is called **opaque**.

**Glass fibres like this are very good at transmitting light. They are used for cable TV, and for connecting up the Internet.**

**◄ CHECKPOINT ►**

What word is the opposite of 'transparent'?

# 8K3 Seeing things

In the centre of your eye is a black hole – the **pupil**. Rays of light pass through the pupil to reach the **retina**, the light-sensitive part of the back of the eye:

- An **image** is formed where rays of light fall on the retina – that's a tiny picture of the thing you are looking at.

- We see **luminous** objects because rays of light travel directly into our eyes.

- We see **non-luminous** objects because they reflect rays into our eyes.

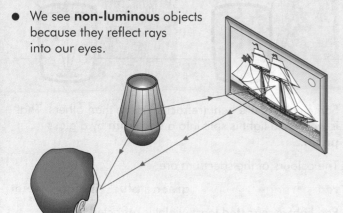

## Ray diagrams

To show how light travels, we draw **ray diagrams**. A ray is a line that shows the path of light. Rays are straight lines, unless they are reflected or something else makes them change direction.

The ray diagram opposite shows how we see the lamp and the picture. If we drew all the rays of light leaving the lamp, and all the rays reflected by the picture, it would be a very complicated and confusing picture. A ray diagram is a simplified way of showing what is going on.

To draw a ray diagram when you are doing an experiment:

- Mark three points on the ray.
- Using a ruler, draw a straight line through the points.

> ### ◄ CHECKPOINT ►
>
> Which is narrower, a ray of light or a beam of light?
>
> _____

# 8K4 Mirrors

When we look into a mirror, we see things by reflected light. The mirror reflects light into our eyes, and we see an **image** of the object in the mirror.

Because the image seems to have been reversed, e.g. left is right, we say that the image is **inverted**.

The **Law of Reflection** allows us to predict where a light ray will go when a mirror reflects it.

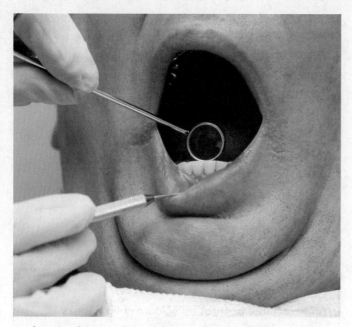

**A dentist has to be able to work while looking at a mirror image of the patient's mouth**

The **normal** is a line drawn at right angles (90°) to the mirror:

- The **angle of incidence** is the angle between the incident ray and the normal.
- The **angle of reflection** is the angle between the reflected ray and the normal.

> angle of incidence = angle of reflection

This is the Law of Reflection.

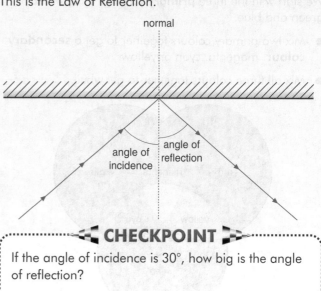

> ### ◄ CHECKPOINT ►
>
> If the angle of incidence is 30°, how big is the angle of reflection?
>
> _____

# 8K5 Refracting light

The diagram shows what happens when a ray of light enters a glass block. You can see that it changes direction twice:

- when it enters the block
- when it leaves the block.

(It's a straight line in between.) This is **refraction**.

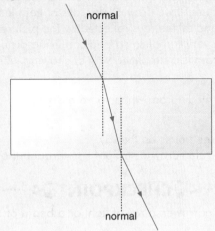

normal

normal

To see what's happening, we draw two normal lines at 90° to the surface of the glass block:

- The ray bends towards the normal when it enters the glass.
- It bends away from the normal when it leaves the glass.

Here's a trick that's explained by refraction. You can't see the coin when the cup is empty. Fill it with water, and the light rays bend to reach your eye.

Some colours are bent (refracted) more than others. That is why white light is split into a **spectrum** by a glass **prism**.

The colours of the spectrum are:

**red** orange yellow **green** **blue** **indigo** **violet**

Red light is refracted least, violet is refracted most.

◀ **CHECKPOINT** ▶

Write the first letter of each colour of the spectrum, in order: R _ _ _ _ _ _

# 8K6 Changing colours

White light is made up of all of the colours of the spectrum. If we mix all these colours together, we get white light.

There's another way to mix colours of light to make white. We start with the three **primary colours** of light: red, green and blue.

- Mix two primary colours together to get a **secondary colour**: magenta, cyan or yellow.
- Mix all three colours together to get white.

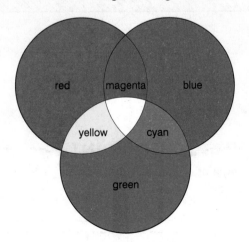

red    magenta    blue

yellow    cyan

green

You can change the colour of white light using a **filter**. A red filter **transmits** (allows through) only red light. It **absorbs** all the other colours that are in the white light.

Things can look different when we shine different colours of light on them:

- A blue flower looks blue in white light, because it reflects blue light and absorbs the other colours.
- It also looks blue in blue light, because it reflects blue light to our eyes.
- It looks black in red light, because it absorbs red light. There is no blue light to reflect.

◀ **CHECKPOINT** ▶

Complete:
An object that absorbs all colours of light looks

_____.

## How light travels (8K1)

**1** Choose words from the list to fill the gaps in the sentences below.

**sound     laser     straight**

A _____ is a source of light.

Light travels in _____ lines.

Light travels more quickly than _____.

**2** Light travels to the Earth from the Sun:

- Distance travelled = 150 000 000 km
- Time taken = 500 s

Use this information to work out the speed at which light travels (speed = distance/time):

_____

_____

**3** Astronomers measure the distances to stars in **light-years**. Find out what is meant by a 'light-year'.

A 'light-year' is _____

_____

## Passing through (8K2)

**1** Draw lines joining words on the left with words on the right to which they are related.

| Transmit | | Opaque |
| Absorb | | Transparent |
| Reflect | | Reflective |
| | | Translucent |

**2 a** In what way is a translucent material similar to a transparent one?

*both let light through*

**b** In what way is a translucent material different from a transparent one?

*transluecent is you can't see through transparent you can see through*

**3 a** A glass of water looks transparent. But is water always transparent? Give an example of a situation where water is not completely transparent:

*when it's frozen*

**b** Is the bottle in the photograph transparent?

*a bit*

## Seeing things (8K3)

**1** How do we describe an object that we can only see by reflected light? _reflective_

**2** In the box on the right, draw a circle around all the luminous objects. Then add two more objects, two of which are luminous (circle them too). _torch_ _Stain_

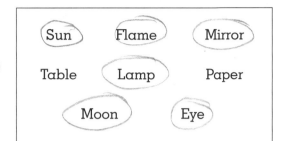

Sun    Flame    Mirror

Table    Lamp    Paper

Moon    Eye

**3** Complete the ray diagram below, to show:

   **a** How the person sees the lamp.

   **b** How the person sees the picture.

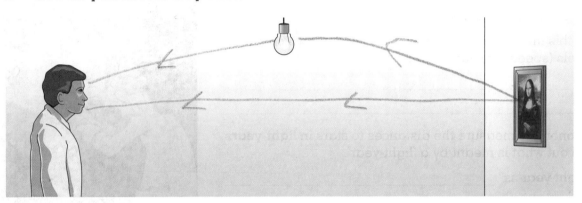

## Mirrors (8K4)

**1** What word means that a mirror image is 'back-to-front'? _____

**2** Complete the table by putting the correct terms in the first column:

| | |
|---|---|
| | A line at right angles to a mirror |
| | A ray of light **before** it strikes a mirror |
| | A ray of light **after** it strikes a mirror |

**3** Complete the Law of Reflection:

angle of _____ = angle of _____

**4** In an experiment, Pip and Reese shone a ray of light from a ray box across a piece of paper. They used a mirror to change the direction of the ray.

They marked the ray before it hit the mirror, and after it had been reflected. Their diagram is shown here. Unfortunately, they forgot to mark the position of the mirror.

   **a** Complete the ray diagram by drawing in the rays. Find the point where they cross.

   **b** Use the Law of Reflection to draw in the position of the mirror.

ray box

1 The diagram shows a ray of light entering a glass block. On the diagram, draw the normal to the surface of the glass, at point X. Mark and label the following:

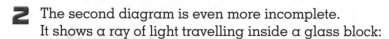

a the incident ray
b the refracted ray
c the angle of incidence
d the angle of refraction.
e Cross out the incorrect word: The angle of incidence is greater/less than the angle of refraction.

2 The second diagram is even more incomplete. It shows a ray of light travelling inside a glass block:

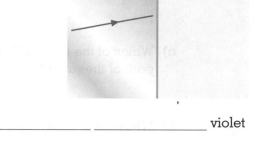

a On the diagram, show how the ray changes direction as it leaves the glass.
Mark and label the rays and angles.

b Cross out the incorrect word:
The angle of incidence is greater/less than the angle of refraction.

3 List the colours of the spectrum, in order:

red _____ _____ _____ _____ _____ violet

4 Which colour of light is refracted least by a glass prism? _____

1 Name the three primary colours of light: _____, _____ and _____.

2 What colour of light is produced by mixing all three primary colours? _____

3 What primary colours of light must be mixed to give:

a magenta? _____ and _____

b cyan? _____ and _____

c yellow? _____ and _____

4 When white light is shone through a green filter:

a What colour of light does it transmit? _____

b What happens to the other colours? They are _____.

5 Look at the photo of the flowers. What colour will they appear:

a in white light? _____

b in blue light? _____

c in yellow light? _____

The diagram shows a ray of light shining onto a flat mirror:

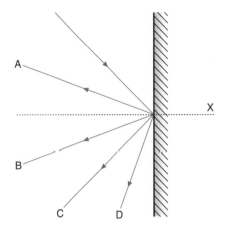

This is a question about the Law of Reflection.

**a)** Which of the rays (A, B, C or D) shows correctly the path of the ray after it has been reflected by the mirror?

_____ (1)

**b)** What name do we give to the line marked X?

_____ (1)

**c)** On the diagram, label the incident and reflected rays. (2)

**d)** Mark two angles which are equal. (1)

 Benson is experimenting with a glass block. He shines a narrow ray of light at it. He tries different positions of the glass block. He draws the results.

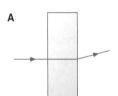 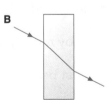

For each of his diagrams, has he shown correctly how the ray could have behaved?

**a)** Diagram A: correct/incorrect (1)

**b)** Diagram B: correct/incorrect (1)

**c)** Diagram C: correct/incorrect (1)

**d)** Diagram D: correct/incorrect (1)

HINT

Things to think about: When does a ray of light bend? Which way does it bend?

**3** Reese shone a ray of white light at a prism, and saw a spectrum on a piece of white card.

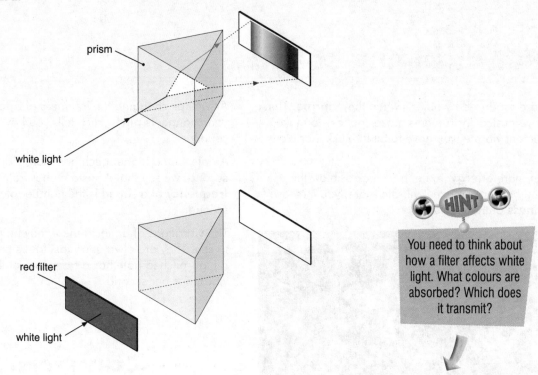

prism

white light

red filter

white light

**HINT**

You need to think about how a filter affects white light. What colours are absorbed? Which does it transmit?

**a)** She put a red filter in the path of the ray of light, as shown in the second diagram. Describe how the appearance of the spectrum would change.

_____ (1)

**b)** She then put the filter between the prism and the white card. What would she see on the card?

_____ (1)

**c)** Explain why Reese used a piece of **white** card in this experiment.

_____

_____ (1)

**HINT**

Think about what she would have seen with other colours of card – black or blue, for example.

# Sound and hearing

## 8L1 Changing sounds

Sounds are produced by objects when they **vibrate**. Think about how musical instruments make sounds. To make an instrument vibrate, you have to hit it, pluck it or blow it.

If you put more **energy** in, i.e. hit, pluck or blow the instrument harder to make it vibrate more, you increase the **loudness** of the sound.

**In a brass band, it's the air inside the instruments that vibrates to make the sounds**

As well as changing the loudness of a sound, a musician can change its **pitch**. Pitch tells you how high or low a sound is.

A note with a higher pitch has more vibrations per second: we say that it has a higher frequency. The **frequency** of a sound is the number of vibrations per second.

The vibrations of an instrument may be too small for us to see. They are always too fast for us to see clearly – a high-pitched note has a frequency of thousands of vibrations per second.

> **◀ CHECKPOINT ▶**
> Cross out the incorrect word:
> If you increase the size of its vibrations, a sound will become louder/higher.

## 8L2 Seeing sounds

You can use an **oscilloscope** to show up the vibrations of a sound.

If the sound gets louder, the trace goes up and down more. We say that its **amplitude** has increased.

If the sound gets higher pitched, more waves are squashed into the same space. Its **frequency** has increased.

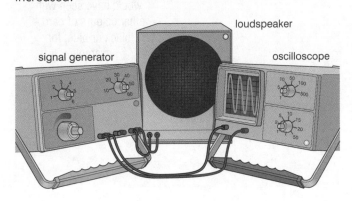

**The amplitude of this sound is changing**

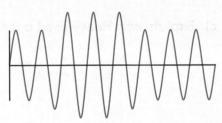

**The frequency of this sound is changing**

> **◀ CHECKPOINT ▶**
> Complete:
> A sound whose amplitude is decreasing is getting
> _____.
> A sound whose frequency is decreasing is getting
> _____.

# 8L3 How sounds travel

Sounds can travel through:

- Solids
- Liquids
- Gases.

They cannot travel through a **vacuum** (an empty space); they need a material medium to travel through.

**There is no air on the Moon, so astronauts must speak to each other using radio transmitters. You can see the aerial sticking out of this astronaut's helmet.**

We say that sound travels as **sound waves**:

- When one particle vibrates, it pushes on its neighbours so that they start to vibrate too.
- Then they push their neighbours, and make them vibrate.
- The vibration travels outwards from the source.

Sounds travel fastest through solids and liquids; they travel more slowly through gases. The particles are closest together in solids and liquids, so the vibrations are passed on quickly from one particle to the next.

**Don't forget!**
The particles only **oscillate** from side to side as a sound vibration travels along. If you shout at someone, the particles of air don't travel from your mouth to their ear!

**◄ CHECKPOINT ►**

Cross out the incorrect words:
Sound travels quickly through air / steel / vacuum.

# 8L4 Hearing things

The frequency of a sound is measured in **hertz** (Hz) or **kilohertz** (kHz):

- 1 Hz means one vibration per second.
- 1 kHz means one thousand vibrations per second.

Young people have a wide **range of hearing**:

- Lowest audible frequency = 20 Hz
- Highest audible frequency = 20 kHz

(These frequencies are only approximate.)

Older people lose the ability to hear such high sounds because the sensitive parts of the ear, particularly the cochlea, gradually deteriorate.

Listening to loud music and other sounds can also reduce the range of hearing.

- Sound waves travel through the air to the **eardrum**.
- The **three small bones** transmit the vibrations to the **cochlea**.
- **Nerves** carry electrical signals to the brain.

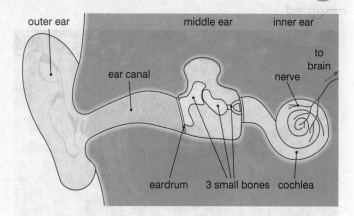

**◄ CHECKPOINT ►**

Which of these frequencies are too high or too low to hear?
Cross them out.

10 Hz    100 Hz    1 kHz    10 kHz    100 kHz

# 8L5 Noise pollution

**Noise** is unwanted sound. We live with sounds around us most of the time and many of these are sounds we like to hear, e.g. people talking, the television, music. But there are also unwanted sounds, e.g. traffic, aircraft, machinery, other people's music. That's **noise pollution**.

- Noise can make it hard to sleep.
- Loud sounds can damage your hearing.
- Constant noise can cause stress.

Ear protection can be worn to reduce the effects of noise, but it's better to reduce the noise.

Sound **insulation** can be fitted to cut down noise. Soft fabrics are good at absorbing sound.

A **sound level meter** measures sounds on the **decibel scale (dB)**.

## The decibel scale of sound

| Sound level (dB) | Example |
| --- | --- |
| 150 | ear bones break |
| 140 | can be painful |
| 130 | the threshold of pain |
| 120 | disco, 1 m from speaker |
| 110 | a pneumatic drill |
| 100 | food blender at 1 m |
| 90 | heavy truck |
| 80 | a door slamming |
| 70 | rush hour traffic |
| 60 | loud conversation |
| 50 | quiet conversation |
| 40 | residential area, night-time |
| 30 | ticking watch at 1 m |
| 20 | leaves rustling in wind |
| 10 | faint whispering |
| 0 | the faintest sound you can hear |

The loudest recorded sound in recent years was when the Mount St Helens' volcano suddenly erupted

Testing a jet engine is a noisy job. This technician will need to wear ear protection.

A baby's cries can reach over 100 dB

## ◄ CHECKPOINT ►

Give a two-word phrase for 'noise':

_____

## Changing sounds (8L1)

**1** Sounds are produced when objects vibrate.
For each of these instruments, name the part that vibrates:

Guitar: _____

Trumpet: _____

Drum: _____

**2** Explain why a piano has so many keys.
Use the word 'pitch' in your answer.

_____

**3** If you hit a drum harder, you give it more _____.

Its vibrations will be _____ and it will sound _____.

**4** Sound A has 200 vibrations each second. Sound B has 400 vibrations each second.

**a** What do we call the number of vibrations per second? _____

**b** Which sound has the higher pitch, A or B? _____

## Seeing sounds (8L2)

**1** What scientific instrument is used to show sounds as traces on a screen?

_____

**2** On the diagram on the right, draw an arrow to show the amplitude of the sound.

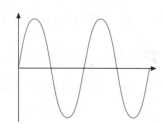

**3** The diagram on the right shows a soft sound.
Add a second trace to the diagram to show
a louder sound with the same frequency.

**4** On the blank grid, draw a trace to represent a sound
whose frequency changes but whose amplitude remains
the same.

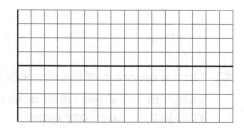

**1** Put a ring around each of the materials in the box that sounds can travel through. Cross out the others.

| Water | Empty space | Wood |
|-------|-------------|------|
| Concrete | Air | Carbon dioxide |
| Vacuum | Glass | Steel |

**2** Use words from the list to complete the sentences, explaining how sound travels through the air:

wave     neighbours     jostle     vibrates     transmitted

A source of sound _____ back and forth.

Its particles _____ the particles of the air.

Each particle causes its _____ to vibrate.

In this way, the vibration is _____ through the air.

We say that a sound _____ has been created.

**3** **a** Complete the description of the experiment with the bell in the vacuum jar:

First, switch on the _____ _____.

Next, use the _____ _____ to

remove the _____ from the jar. Now

there is a _____ in the jar.

**b** What do you observe? _____

_____

**c** What does this experiment show? _____

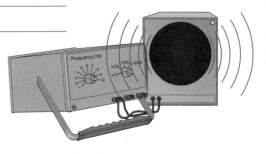

to vacuum pump

power supply

**1** **a** What quantity is measured in hertz? _____

**b** What is meant by 1 Hz? _____

**c** What is meant by 20 kHz? _____

**2** The drawing shows an experiment to investigate hearing:

**a** Which part of the apparatus produces the sound?

_____

**b** Describe how to increase the frequency of the sound:

_____

**c** What happens when the frequency is increased beyond 20 kHz?

_____

**d** What is the normal range of hearing for a child?

From _____ to _____.

**3** On the diagram of the ear, label the following:

eardrum     three small bones
cochlea     auditory nerve

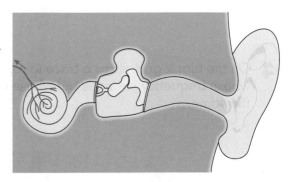

1   Use words from the list to complete the sentences below:

**insulation    decibel    stress    hearing    sound**

Noise is unwanted _____.

It is measured on the _____ scale.

Noise can cause _____; it can also damage your _____.

Noise can be reduced by the use of _____.

2   a   In the box below, draw a noisy scene.

    b   Label all of the sources of noise.

    c   Underneath, give some ideas about how the amount of noise could be reduced.

_____

_____

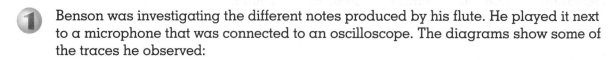

**1** Benson was investigating the different notes produced by his flute. He played it next to a microphone that was connected to an oscilloscope. The diagrams show some of the traces he observed:

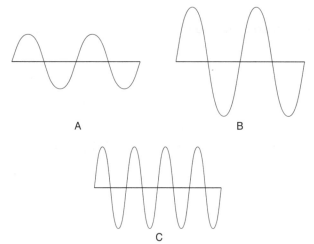

A                    B

C

HINT

Think carefully about the amplitude and frequency of each trace.

a) Which diagram represents a loud, high-pitched note? _____ (1)

b) Which diagram represents a loud, low-pitched note? _____ (1)

c) Which diagram represents a soft, low-pitched note? _____ (1)

**2** Pip and Molly were investigating different materials, to see which was the best absorber of sound. Molly put squares of material over her ears, and Pip played her trumpet nearby. Molly judged which material let through the least sound.

HINT

Look at the picture to see how they carried out the experiment.

a) List **three** things that Molly and Pip should keep the same for this to be a fair test.

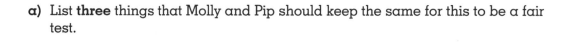

_____

_____

_____ (3)

b) Why would it be better to use a signal generator and a loudspeaker to produce the sound, rather than a trumpet?

_____ (1)

**3** Small children have to learn that it is dangerous to poke things into their ears, as this can damage the eardrum.

a) Describe how the eardrum moves when a sound enters the ear:

_____

(1)

b) How does its motion change if the sound gets louder?

_____

_____

(1)

c) How does its motion change if the pitch of the sound gets lower?

_____

_____

(1)

**HINT**

These parts of the question test whether you understand about the **amplitude** and **frequency** of a sound.

# Acknowledgements

## Picture Acknowledgements

Bettman/Corbis: 35; Charles D. Winters/Science Photo Library: 48b; Corel 8 (NT): 85; Corel 12 (NT): 28a; Corel 16 (NT): 28b; Corel 244 (NT): 27; Corel 269 (NT): 5; Corel 270 (NT): 18a; Corel 340 (NT): 48a; Corel 459 (NT): 2; Corel 493 (NT): 21; Corel 587 (NT): 56b, 59b; Corel 671 (NT): 90b; Corel 689 (NT): 50c; Digital Stock 12 (NT); 18b; Digital Vision 5 (NT): 22; Digital Vision 6 (NT): 89; Digital Vision PB (NT): 88; 90c; Dr. B. Booth/GeoScience Features Picture Library: 50a, b, 56a, 59c; George Bernard/Science Photo Library: 57b; Dirk Wiersma/Science Photo Library: 57c; GeoScience Features Picture Library: 58a, b; Kaj R. Svensson/Science Photo Library: 48c; Photodisc 4 (NT): 90a; Photodisc 6 (NT): 66; Photodisc 18 (NT): 65, 81; Photodisc 24 (NT): 91; Photodisc 40 (NT): 20; Photodisc 44 (NT): 57a; Photodisc 54 (NT): 64, 73b, 80; Photodisc 67 (NT): 83, 9; Photodisc 72 (NT): 72, 73a; RIDA/GeoScience Features Picture Library: 59a; Stephen Frink/Digital Vision LU (NT): 12.

Every effort has been made to trace the copyright holders, but if any have been overlooked, the publishers will be pleased to make the necessary arrangements at the first opportunity.

Picture research by Stuart Sweatmore.

## How to order Scientifica

| Title | ISBN | Price £ | IC/A please tick | ✔ | Firm Order Qty | £ |
|---|---|---|---|---|---|---|
| **Student Book (Levels 4–7)** | 0 7487 7988 4 | 10.00 | IC | | | |
| **Student Book Essentials (Levels 3–6)** | 0 7487 7989 2 | 10.00 | IC | | | |
| **Teacher Book 8 (Levels 4–7)** | 0 7487 7992 2 | 40.00 | A | | | |
| **Teacher Book 8 Essentials (Levels 3–6)** | 0 7487 7995 7 | 40.00 | A | | | |
| **Teacher Resource Pack 8** | 0 7487 8026 2 | 100.00 | A | | | |
| **ICT Power Pack 8 (eligible for eLC funding)** | 0 7487 8027 0 | 250.00 + | A | | | |
| **Assessment Pack 8** | 0 7487 8030 0 | 125.00 | A | | | |
| **Special Resource Pack 8** | 0 7487 9202 3 | 150.00 | A | | | |
| **Workbook 8** | 0 7487 9185 X | 3.50 | IC | | | |
| **Scientifica Presents The Amazement Park 8** | 0 7487 9014 4 | 6.50 | IC | | | |
| * Inspection or approval copies will be delivered carriage free to educational establishments. £2.95 post and packing will be charged on all firm orders. | | | **Post and Packaging** | | | £2.95* |
| | | | **Total** | | | |

**Year 7** material is now available     **Year 9** material available from **Summer 2005**

Prices valid until 31–12–05 – but are subject to change without notice

**IC** – Inspection Copy     **A** – On Approval     **+** – Plus VAT

Inspection/Approval copies are available to recognised educational establishments for evaluation free of charge for up to 30 days.

For terms and conditions please contact Customer Services on 01242 267287.

To request an inspection/approval copy, or to place a firm order, please complete and return this card, quoting an Official Order Number if required.

Official Order Number _____     To ensure your order is not delayed please check whether this is required by your establishment.

Name _____

Position _____

School/College _____

School/College address _____

_____

Postcode _____     Email _____

Telephone _____     Fax _____

Please send me a Science Catalogue ☐     Please send me an Electronic Catalogue ☐     SFSV

**To contact your local educational representative call our dedicated support line on 01242 267284**

## Please photocopy and send back to:

**Customer Services, Nelson Thornes,**
**FREEPOST SWC0504, Cheltenham, Glos. GL53 7ZZ**

t: 01242 267273   f: 01242 253695   e: science@nelsonthornes.com   w: www.nelsonthornes.com